ИЗДАТЕЛЬСТВО • МИР •

И. Е. Иродов

ОСНОВНЫЕ ЗАКОНЫ МЕХАНИКИ

Издательство «Высшая школа»
Москва

I.E. IRODOV

FUNDAMENTAL LAWS OF MECHANICS

Translated from the Russian
by YURI ATANOV

Mir Publishers Moscow

CBS

CBS Publishers & Distributors Pvt. Ltd.

New Delhi • Bengaluru • Chennai • Kochi • Kolkata • Mumbai
Hyderabad • Nagpur • Patna • Pune • Vijayawada

First published 1980
Revised from the 1978 Russian edition

На английском языке

ISBN: 81-239-0304-9

First Indian Edition: 1994
Reprint: 2001, 2002, 2004

This edition has been published in India by arrangement with
Mir Publishers, Moscow

Published by:
Satish Kumar Jain for CBS Publishers & Distributors Pvt. Ltd.,
4819/XI Prahlad Street, 24 Ansari Road, Daryaganj, New Delhi - 110002
delhi@cbspd.com, cbspubs@airtelmail.in • www.cbspd.com
Ph.: 23289259, 23266861, 23266867 • Fax: 011-23243014

Corporate Office: 204 FIE, Industrial Area, Patparganj, Delhi - 110 092
Ph: 49344934 • Fax: 011-49344935
E-mail: publishing@cbspd.com • publicity@cbspd.com

Branches:
- ***Bengaluru:*** 2975, 17th Cross, K.R. Road, Bansankari 2nd Stage, Bengaluru - 70 • Ph: +91-80-26771678/79 • Fax: +91-80-26771680 E-mail: cbsbng@gmail.com, bangalore@cbspd.com
- ***Chennai:*** No. 7, Subbaraya Street, Shenoy Nagar, Chennai - 600030 Ph: +91-44-26681266, 26680620 • Fax: +91-44-42032115 E-mail: chennai@cbspd.com
- ***Kochi:*** Ashana House, 39/1904, A.M. Thomas Road, Valanjambalam, Ernakulum, Kochi • Ph: +91-484-4059061-65 Fax: +91-484-4059065 • E-mail: cochin@cbspd.com
- ***Kolkata:*** 6-B, Ground Floor, Rameshwar Shaw Road, Kolkata - 700014 Ph: +91-33-22891126/7/8 • E-mail: kolkata@cbspd.com
- ***Mumbai:*** 83-C, Dr. E. Moses Road, Worli, Mumbai - 400018 Ph: +91-9833017933, 022-24902340/41 • E-mail: mumbai@cbspd.com

Representatives:
- Hyderabad: 0-9885175004
- Nagpur: 0-9021734563
- Patna: 0-9334159340
- Pune: 0-9623451994
- Vijayawada: 0-9000660880

Printed at: Neekunj Print Process, Delhi

CONTENTS

Preface 7
Notation 9
Introduction 11

PART ONE
CLASSICAL MECHANICS

Chapter 1. Essentials of Kinematics 14
§ 1.1. Kinematics of a Point 14
§ 1.2. Kinematics of a Solid 21
§ 1.3. Transformation of Velocity and Acceleration on Transition to Another Reference Frame 30
Problems to Chapter 1 34

Chapter 2. The Basic Equation of Dynamics 41
§ 2.1. Inertial Reference Frames 41
§ 2.2. The Fundamental Laws of Newtonian Dynamics 44
§ 2.3. Laws of Forces 50
§ 2.4. The Fundamental Equation of Dynamics 52
§ 2.5. Non-inertial Reference Frames. Inertial Forces 56
Problems to Chapter 2 61

Chapter 3. Energy Conservation Law 72
§ 3.1. On Conservation Laws 72
§ 3.2. Work and Power 74
§ 3.3. Potential Field of Forces 79
§ 3.4. Mechanical Energy of a Particle in a Field . . . 90
§ 3.5. The Energy Conservation Law for a System . . . 94
Problems to Chapter 3 104

Chapter 4. The Law of Conservation of Momentum 114
§ 4.1. Momentum. The Law of Its Conservation 114
§ 4.2. Centre of Inertia. The *C* Frame 120
§ 4.3. Collision of Two Particles 126
§ 4.4. Motion of a Body with Variable Mass 136
Problems to Chapter 4 139

Chapter 5. The Law of Conservation of Angular Momentum 147
§ 5.1. Angular Momentum of a Particle. Moment of Force 147
§ 5.2. The Law of Conservation of Angular Momentum 154
§ 5.3. Internal Angular Momentum 160
§ 5.4. Dynamics of a Solid 164
Problems to Chapter 5 178

PART TWO
RELATIVISTIC MECHANICS

Chapter 6. Kinematics in the Special Theory of Relativity . 189

§ 6.1. Introduction 189
§ 6.2. Einstein's Postulates 194
§ 6.3. Dilation of Time and Contraction of Length 198
§ 6.4. Lorentz Transformation 207
§ 6.5. Consequences of Lorentz Transformation 210
§ 6.6. Geometric Description of Lorentz Transformation 217
Problems to Chapter 6 221

Chapter 7. Relativistic Dynamics 227

§ 7.1. Relativistic Momentum 227
§ 7.2. Fundamental Equation of Relativistic Dynamics 231
§ 7.3. Mass-Energy Relation 233
§ 7.4. Relation Between Energy and Momentum of a Particle 238
§ 7.5. System of Relativistic Particles 242
Problems to Chapter 7 249

Appendices 257

1. Motion of a Point in Polar Coordinates 257
2. On Keplerian Motion 258
3. Demonstration of Steiner's Theorem 260
4. Greek Alphabet 262
5. Some Formulae of Algebra and Trigonometry . . . 262
6. Table of Derivatives and Integrals 263
7. Some Facts About Vectors 264
8. Units of Mechanical Quantities in the SI and CGS Systems 266
9. Decimal Prefixes for the Names of Units 266
10. Some Extrasystem Units 267
11. Astronomic Quantities 267
12. Fundamental Constants 267

Index 269

PREFACE

The objective of the book is to draw the readers' attention to the *basic* laws of mechanics, that is, to the laws of motion and to the laws of conservation of energy, momentum and angular momentum, as well as to show *how* these laws are to be applied in solving various specific problems. At the same time the author has excluded all things of minor importance in order to concentrate on the questions which are the hardest to comprehend.

The book consists of two parts: (1) classical mechanics and (2) relativistic mechanics. In the first part the laws of mechanics are treated in the Newtonian approximation, i.e. when motion velocities are much less than the velocity of light, while in the second part of the book velocities comparable to that of light are considered.

Each chapter opens with a theoretical essay followed by a number of the most instructive and interesting examples and problems, with solutions provided. There are about 80 problems altogether; being closely associated with the introductory text, they develop and supplement it and therefore their examination is of equal importance.

A few corrections and refinements have been made in the present edition to stress the physical essence of the problems studied. This holds true primarily for Newton's second law and the conservation laws. Some new examples and problems have been provided.

The book is intended for first-year students of physics but can also be useful to senior students and lecturers.

I. E. Irodov

NOTATION

Vectors are designated by roman bold-face type (e.g. $\mathbf{r}$, $\mathbf{F}$); the same italicized letters (r, F) designate the norm of a vector.

Mean values are indicated by crotchets $\langle\ \rangle$, e.g. $\langle \mathbf{v} \rangle$, $\langle N \rangle$.

The symbols Δ, d, δ (when put in front of a quantity) signify:

Δ, a finite increment of a quantity, i.e. a difference between its final and initial values, e.g. $\Delta\mathbf{r} = \mathbf{r}_2 - \mathbf{r}_1$, $\Delta U = U_2 - U_1$;
d, a differential (an infinitesimal increment), e.g. $d\mathbf{r}$, dU;
δ, an elementary value of a quantity, e.g. δA is an elementary work.

Unit vectors:

$\mathbf{i}$, $\mathbf{j}$, $\mathbf{k}$ are unit vectors of the Cartesian coordinates x, y, z;
$\mathbf{e}_\rho$, $\mathbf{e}_\varphi$, $\mathbf{e}_z$ are unit vectors of the cylindrical coordinates ρ, φ, z;
$\mathbf{n}$, $\boldsymbol{\tau}$ are unit vectors of a normal and a tangent to a path.

Reference frames are denoted by the italic letters K, K' and C.

The C frame is a reference frame fixed to the centre of inertia and translating relative to inertial frames. All quantities in the C frame are marked with a tilde, e.g. $\tilde{p}$, $\tilde{E}$.

A, work,
c, velocity of light *in vacuo*,
E, total mechanical energy, the total energy,
$\mathbf{E}$, electric field strength,
e, elementary electric charge,
$\mathbf{F}$, force,
$\mathbf{G}$, field strength,
$\mathbf{g}$, free fall acceleration,
I, moment of inertia,
$\mathbf{L}$, angular momentum with respect to a point,
L_z, angular momentum with respect to an axis,

l, arc coordinate, the arm of a vector,
$\mathbf{M}$, moment of a force with respect to a point,
M_z, moment of a force with respect to an axis,
m, mass, relativistic mass, m_0 rest mass,
N, power,
$\mathbf{p}$, momentum,
q, electric charge,
$\mathbf{r}$, radius vector,
s, path, interval,
t, time,
T, kinetic energy,
U, potential energy,
$\mathbf{v}$, velocity of a point or a particle,
$\mathbf{w}$, acceleration of a point or a particle,
$\boldsymbol{\beta}$, angular acceleration,
β, velocity expressed in units of the velocity of light,
γ, gravitational constant, the Lorentz factor,
ε, energy of a photon,
$\varkappa$, elastic (quasi-elastic) force constant,
μ, reduced mass,
ρ, curvature radius, radius vector of the shortest distance to an axis, density,
φ, azimuth angle, potential,
$\boldsymbol{\omega}$, angular velocity,
Ω, solid angle.

INTRODUCTION

Mechanics is a branch of physics treating the simplest form of motion of matter, mechanical motion, that is, the motion of bodies in space and time. The occurrence of mechanical phenomena in space and time can be seen in any mechanical law involving, explicitly or implicitly, space-time relations, i.e. distances and time intervals.

The position of a body in space can be determined only with respect to other bodies. The same is true for the motion of a body, i.e. for the change in its position over time. The body (or the system of mutually immobile bodies) serving to define the position of a particular body is identified as the reference body.

For practical purposes, a certain coordinate system, e.g. the Cartesian system, is fixed to the reference body whenever motion is described. The coordinates of a body permit its position in space to be established. Next, motion occurs not only in space but also in time, and therefore the description of the motion presupposes time measurements as well. This is done by means of a clock of one or another type.

A reference body to which coordinates are fixed and mutually synchronized clocks form the so-called *reference frame*. The notion of a reference frame is fundamental in physics. A space-time description of motion based on distances and time intervals is possible only when a definite reference frame is chosen.

Space and time by themselves are also *physical* objects, just as any others, even though immeasurably more important. The properties of space and time can be investigated by observing bodies moving in them. By studying the character of the motion of bodies we determine the properties of space and time.

Experience shows that as long as the velocities of bodies are small in comparison with the velocity of light, linear scales and time intervals remain *invariable* on transition from one reference frame to another, i.e. they do not depend on the choice of a reference frame. This fact finds expression in the Newtonian concepts of absolute space and time. Mechanics treating the motion of bodies in such cases is referred to as classical.

When we pass to velocities comparable to that of light, it becomes obvious that the character of the motion of bodies changes radically. Linear scales and time intervals become *dependent* on the choice of a reference frame and are different in different reference frames. Mechanics based on these concepts is referred to as *relativistic*. Naturally, relativistic mechanics is more general and becomes classical in the case of small velocities.

The actual motion characteristics of bodies are so complex that to investigate them we have to neglect all insignificant factors, otherwise the problem would get so complicated as to render it practically insoluble. For this purpose notions (or abstractions) are employed whose application depends on the specific nature of the problem in question and on the accuracy of the result that we expect to get. A particularly important role is played by the notions of a mass point and of a perfectly rigid body.

A *mass point*, or, briefly, a *particle*, is a body whose dimensions can be neglected under the conditions of a given problem. It is clear that the same body can be treated as a mass point in some cases and as an extended object in others.

A *perfectly rigid body*, or, briefly, a *solid*, is a system of mass points separated by distances which do not vary during its motion. A real body can be treated as a perfectly rigid one provided its deformations are negligible under the conditions of the problem considered.

Mechanics tackles two fundamental problems:

1. The investigation of various motions and the generalization of the results obtained in the form of laws of mo-

tion, i.e. laws that can be employed in predicting the character of motion in each specific case.

2. The search for general properties that are typical of any system regardless of the specific interactions between the bodies of the system.

The solution of the first problem ended up with the so-called dynamic laws established by Newton and Einstein, while the solution of the second problem resulted in the discovery of the laws of conservation for such fundamental quantities as energy, momentum and angular momentum.

The dynamic laws and the laws of conservation of energy, momentum and angular momentum represent the basic laws of mechanics. The investigation of these laws constitutes the subject matter of this book.

PART ONE

CLASSICAL MECHANICS

CHAPTER 1

ESSENTIALS OF KINEMATICS

Kinematics is the subdivision of mechanics treating ways of describing motion regardless of the causes inducing it. Three problems will be considered in this chapter: kinematics of a point, kinematics of a solid, and the transformation of velocity and acceleration on transition from one reference frame to another.

§ 1.1. Kinematics of a Point

There are three ways to describe the motion of a point: the first employs vectors, the second coordinates, and the third is referred to as natural. Let us examine them in order.

The vector method. With this method the location of a given point A is defined by a radius vector $\mathbf{r}$ drawn from a certain stationary point O of a chosen reference frame to that point A. The motion of the point A makes its radius vector vary in the general case both in magnitude and in direction, i.e. the radius vector $\mathbf{r}$ depends on time t. The locus of the end points of the radius vector $\mathbf{r}$ is referred to as the *path* of the point A.

Let us introduce the notion of the *velocity* of a point. Suppose the point A travels from point *1* to point *2* in the time interval Δt (Fig. 1). It is seen from the figure that the *displacement vector* $\Delta\mathbf{r}$ of the point A represents the increment of the radius vector $\mathbf{r}$ in the time Δt: $\Delta\mathbf{r} = \mathbf{r}_2 -$

— $\mathbf{r}_1$. The ratio $\Delta\mathbf{r}/\Delta t$ is called the *mean velocity vector* $\langle\mathbf{v}\rangle$ during the time interval Δt. The direction of the vector $\langle\mathbf{v}\rangle$ coincides with that of $\Delta\mathbf{r}$. Now let us define the velocity vector $\mathbf{v}$ of the point at a given moment of time as the limit of the ratio $\Delta\mathbf{r}/\Delta t$ as $\Delta t \to 0$, i.e.

$$\mathbf{v} = \lim_{\Delta t \to 0} \frac{\Delta \mathbf{r}}{\Delta t} = \frac{d\mathbf{r}}{dt}. \tag{1.1}$$

This means that the velocity vector $\mathbf{v}$ of the point at a given moment of time is equal to the derivative of the radius vector $\mathbf{r}$ with respect to time, and its direction, like that of

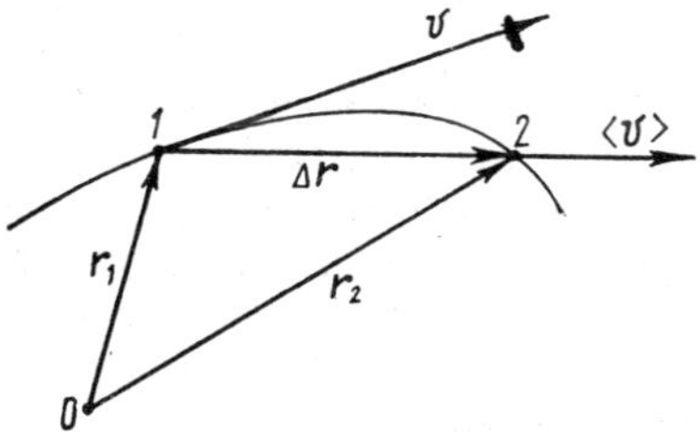

Fig. 1

the vector $d\mathbf{r}$, along the tangent to the path at a given point coincides with the direction of motion of the point A. The modulus of the vector $\mathbf{v}$ is equal to*

$$v = |\mathbf{v}| = |d\mathbf{r}/dt|.$$

The motion of a point is also characterized by *acceleration.* The acceleration vector $\mathbf{w}$ defines the rate at which the velocity vector of a point varies with time:

$$\mathbf{w} = d\mathbf{v}/dt, \tag{1.2}$$

i.e. it is equal to the derivative of the velocity vector with respect to time. The direction of the vector $\mathbf{w}$ coincides with the direction of the vector $d\mathbf{v}$ which is the increment of the vector $\mathbf{v}$ during the time interval dt. The modulus of the

* Note that in the general case $|d\mathbf{r}| \neq dr$, where r is the modulus of the radius vector $\mathbf{r}$, and $v \neq dr/dt$. For example, when $\mathbf{r}$ changes only in direction, that is the point moves in a circle, then $r = \text{const}$, $dr = 0$, but $|d\mathbf{r}| \neq 0$.

vector **w** is defined in much the same way as that of the vector **v**.

Example. The radius vector of a point depends on time t as

$$\mathbf{r} = \mathbf{a}t + \mathbf{b}t^2/2,$$

where **a** and **b** are constant vectors. Let us find the velocity **v** of the point and its acceleration **w**:

$$\mathbf{v} = d\mathbf{r}/dt = \mathbf{a} + \mathbf{b}t, \quad \mathbf{w} = d\mathbf{v}/dt = \mathbf{b} = \text{const.}$$

The modulus of the velocity vector

$$v = \sqrt{\mathbf{v}^2} = \sqrt{a^2 + 2\mathbf{ab}t + b^2t^2}.$$

Thus, knowing the function $\mathbf{r}(t)$, one can find the velocity **v** of a point and its acceleration **w** at any moment of time.

Here the reverse problem arises: can we find $\mathbf{v}(t)$ and $\mathbf{r}(t)$ if the time dependence of the acceleration $\mathbf{w}(t)$ is known?

It turns out that the dependence $\mathbf{w}(t)$ is not sufficient to get a single-valued solution of this problem; one needs also to know the so-called *initial conditions*, namely, the velocity $\mathbf{v}_0$ of the point and its radius vector $\mathbf{r}_0$ at a certain initial moment $t = 0$. To make sure, let us examine the simple case when the acceleration of the point remains constant in the course of time.

First, let us determine the velocity $\mathbf{v}(t)$ of the point. In accordance with Eq. (1.2) the elementary velocity increment during the time interval dt is equal to $d\mathbf{v} = \mathbf{w}\,dt$. Integrating this relation with respect to time between $t = 0$ and t, we obtain the velocity vector increment during this interval:

$$\Delta\mathbf{v} = \int_0^t \mathbf{w}\,dt = \mathbf{w}t.$$

However, the quantity $\Delta\mathbf{v}$ is not the required velocity **v**. To find **v**, we must know the velocity $\mathbf{v}_0$ at the initial moment of time. Then $\mathbf{v} = \mathbf{v}_0 + \Delta\mathbf{v}$, or

$$\mathbf{v} = \mathbf{v}_0 + \mathbf{w}t.$$

The radius vector $\mathbf{r}(t)$ of the point is found in a similar manner. According to Eq. (1.1) the elementary increment of the radius vector during the time interval dt is $d\mathbf{r} = \mathbf{v}\,dt$.

Integrating this relation with respect to the function $\mathbf{v}(t)$, we obtain the increment of the radius vector during the interval from $t = 0$ to t:

$$\Delta\mathbf{r} = \int_0^t \mathbf{v}(t)\,dt = \mathbf{v}_0 t + \mathbf{w}t^2/2.$$

To find the radius vector $\mathbf{r}(t)$, the location $\mathbf{r}_0$ of the point at the initial moment of time must be known. Then $\mathbf{r} = \mathbf{r}_0 + \Delta\mathbf{r}$, or

$$\mathbf{r} = \mathbf{r}_0 + \mathbf{v}_0 t + \mathbf{w}t^2/2.$$

Let us consider, for example, the motion of a stone thrown with the initial velocity $\mathbf{v}_0$ at an angle to the horizontal. Assuming the stone to move with the constant acceleration $\mathbf{w} = \mathbf{g}$, its location relative to the point $\mathbf{r}_0 = 0$ from which the stone was thrown is defined by the radius vector

$$\mathbf{r} = \mathbf{v}_0 t + \mathbf{g}t^2/2,$$

i.e. in this case $\mathbf{r}$ represents the sum of two vectors as shown in Fig. 2.

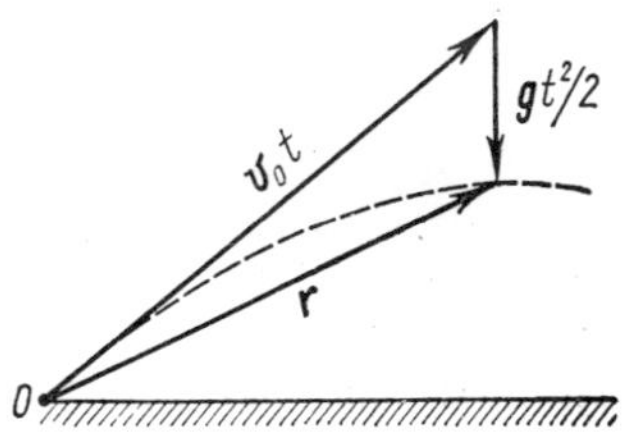

Fig. 2

Thus, the complete solution of the problem of a moving point, that is, the determination of its vcleoity $\mathbf{v}$ and its location $\mathbf{r}$ as functions of time, requires knowing not only the dependence $\mathbf{w}(t)$, but also the initial conditions, i.e. the velocity $\mathbf{v}_0$ and the location $\mathbf{r}_0$ of the point at the initial moment of time.

The method of coordinates. In this method a certain coordinate system (Cartesian, oblique-angled or curvilinear) is fixed to a chosen reference body. The choice of a coordinate system is stipulated by various considerations: the character or the symmetry of the problem, the formulation of the problem, the quest for a simpler solution. We shall confine ourselves here* to Cartesian coordinates x, y, z.

* The motion of a point in polar coordinates is considered in Appendix 1.

Let us write the projections of the radius vector **r** (t) on the axes x, y, z to characterize the position of the point in question relative to the origin O at the moment t:

$$x = x(t); \quad y = y(t); \quad z = z(t).$$

Knowing the dependence of these coordinates on time, that is, *the law of motion* of the point, we can find the position of the point at any moment of time, as well as its velocity and acceleration. Indeed, from Eqs. (1.1) and (1.2) we can easily obtain the formulae defining the projections of the velocity vector and the acceleration vector on the x axis:

$$v_x = dx/dt, \tag{1.3}$$

where dx is the projection of the displacement vector $d\mathbf{r}$ on the x axis;

$$w_x = dv_x/dt = d^2x/dt^2, \tag{1.4}$$

where dv_x is the projection of the velocity increment vector $d\mathbf{v}$ on the x axis. Similar relations are obtained for y and z projections of the respective vectors. It is seen from these formulae that the velocity and acceleration vector projections are equal respectively to the first and second time derivatives of the coordinates.

Thus, the functions $x(t)$, $y(t)$, $z(t)$, in essence, completely define the motion of a point. Knowing them, one can find not only the position of a point, but also the projections of its velocity and acceleration, and, consequently, the magnitude and direction of vectors **v** and **w** at any moment of time. For example, the modulus of the velocity vector

$$v = \sqrt{v_x^2 + v_y^2 + v_z^2};$$

the direction of the vector **v** is defined by the directional cosines as follows:

$$\cos\alpha = v_x/v; \quad \cos\beta = v_y/v; \quad \cos\gamma = v_z/v,$$

where α, β, γ are the angles formed by the vector **v** with the axes x, y, z respectively. Similar formulae define the magnitude and direction of the acceleration vector.

Besides, some more questions can be solved: one can determine the path of a point, the dependence of the distance covered on time, the dependence of the velocity on the position of a point etc.

The reverse problem, that is, the determination of the velocity and the law of motion of a point from a given acceleration, is solved, as in the vector method, by integration (in this case, integration of acceleration projections with respect to time); this problem also has a single-valued solution provided that in addition to the acceleration the initial conditions are also available, i.e. velocity projections and the coordinates of a point at the initial moment.

The "natural" method. This method is employed when the path of a point is known in advance. The location of a point A is defined by the *arc coordinate* l, that is, the distance from the chosen origin O measured along the path (Fig. 3). In so doing, the positive direction of the coordinate l is adopted at will (e.g. as shown by an arrow in the figure).

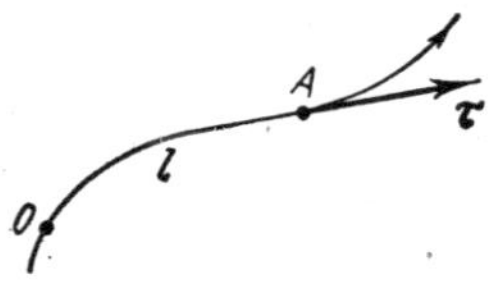

Fig. 3

The motion of a point is determined provided we know its path, the origin O, the positive direction of the arc coordinate l and the law of motion of the point, i.e. the function $l(t)$.

Velocity of a point. Let us introduce the unit vector $\boldsymbol{\tau}$ fixed to the moving point A and oriented along a tangent to the path in the direction of growing values of the arc coordinate l (Fig. 3). It is obvious that $\boldsymbol{\tau}$ is a variable vector since it depends on l. The velocity vector $\mathbf{v}$ of the point A is oriented along a tangent to the path and therefore can be represented as follows:

$$\boxed{\mathbf{v} = v_\tau \boldsymbol{\tau},} \tag{1.5}$$

where $v_\tau = dl/dt$ is the projection of the vector $\mathbf{v}$ on the direction of the vector $\boldsymbol{\tau}$, with v_τ being an algebraic quantity. Besides, it is obvious that

$$|v_\tau| = |\mathbf{v}| = v.$$

Acceleration of a point. Let us differentiate Eq. (1.5) with respect to time:

$$\mathbf{w} = \frac{d\mathbf{v}}{dt} = \frac{dv_\tau}{dt}\boldsymbol{\tau} + v_\tau \frac{d\boldsymbol{\tau}}{dt}. \tag{1.6}$$

2*

Then transform the last term of this expression:

$$v_\tau \frac{d\boldsymbol{\tau}}{dt} = v_\tau \frac{d\boldsymbol{\tau}}{dl}\frac{dl}{dt} = v_\tau^2 \frac{d\boldsymbol{\tau}}{dl} = v^2 \frac{d\boldsymbol{\tau}}{dl}. \tag{1.7}$$

Let us determine the increment of the vector $\boldsymbol{\tau}$ in the interval dl (Fig. 4). It can be strictly shown that when point *2* approaches point *1*, the segment of the path between them tends to turn into an arc of a circle with centre at some

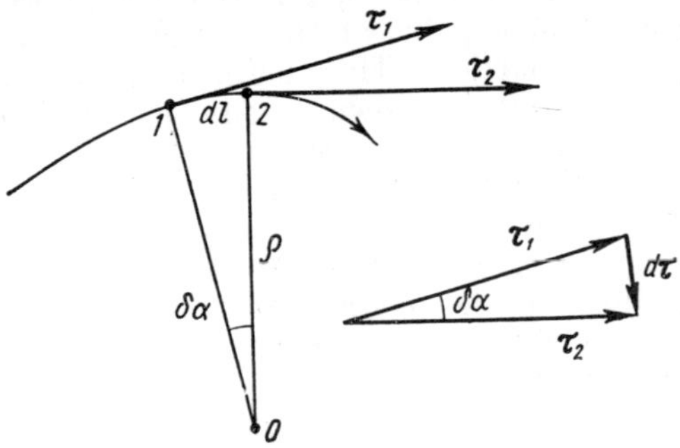

Fig. 4

point O. The point O is referred to as the centre of curvature of the path at the given point, and the radius ρ of the corresponding circle as the radius of curvature of the path at the same point.

It is seen from Fig. 4 that the angle $\delta\alpha = |dl|/\rho = |d\boldsymbol{\tau}|/1$, whence

$$|d\boldsymbol{\tau}/dl| = 1/\rho;$$

at the same time, if $dl \to 0$, then $d\boldsymbol{\tau} \perp \boldsymbol{\tau}$. Introducing a unit vector $\mathbf{n}$ of the normal to the path at point *1* directed toward the centre of curvature, we write the last equality in a vector form:

$$d\boldsymbol{\tau}/dl = \mathbf{n}/\rho. \tag{1.8}$$

Now let us substitute Eq. (1.8) into Eq. (1.7) and then the expression obtained into Eq. (1.6). Finally we get

$$\boxed{\mathbf{w} = \frac{dv_\tau}{dt}\boldsymbol{\tau} + \frac{v^2}{\rho}\mathbf{n}.} \tag{1.9}$$

Here the first term is called the *tangential acceleration* $\mathbf{w}_\tau$ and the second one, the *normal* (centripetal) *acceleration* $\mathbf{w}_n$:

$$\mathbf{w}_\tau = (dv_\tau/dt)\,\boldsymbol{\tau}; \qquad \mathbf{w}_n = (v^2/\rho)\,\mathbf{n}. \tag{1.10}$$

Thus, the total acceleration $\mathbf{w}$ of a point can be represented as the sum of the tangential $\mathbf{w}_\tau$ and normal $\mathbf{w}_n$ accelerations.

The magnitude of the total acceleration of a point is

$$w = \sqrt{w_\tau^2 + w_n^2} = \sqrt{(dv/dt)^2 + (v^2/\rho)^2}.$$

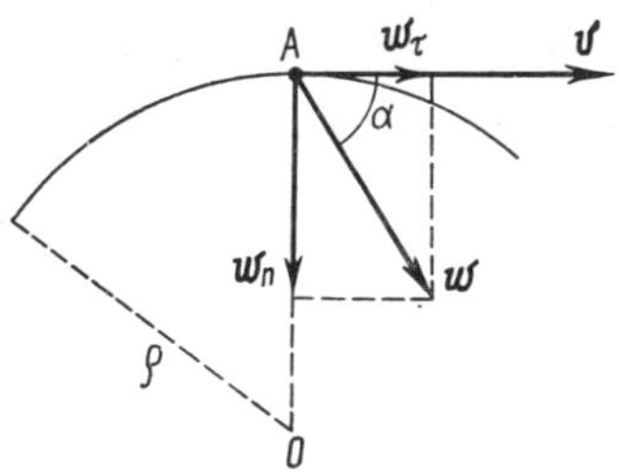

Fig. 5

Example. Point A travels along an arc of a circle of radius ρ (Fig. 5). Its velocity depends on the arc coordinate l as $v = a\sqrt{l}$ where a is a constant. Let us calculate the angle α between the vectors of the total acceleration and of the velocity of the point as a function of the coordinate l.

It is seen from Fig. 5 that the angle α can be found by means of the formula $\tan\alpha = w_n/w_\tau$. Let us find w_n and w_τ:

$$w_n = \frac{v^2}{\rho} = \frac{a^2 l}{\rho}; \quad w_\tau = \frac{dv_\tau}{dt} = \frac{dv_\tau}{dl}\frac{dl}{dt} = \frac{a}{2\sqrt{l}}\,a\sqrt{l} = \frac{a^2}{2}.$$

Whence $\tan\alpha = 2l/\rho$.

§ 1.2. Kinematics of a Solid

Being important by itself, the theory of motion of a solid is also essential in another respect. It is well known that a reference frame used for describing various kinds of motion in space and time can be fixed to a solid. Therefore, the study of motion of solids is actually equivalent to the study of motion of corresponding reference frames. The results to be obtained in this section will be repeatedly used hereafter.

Five kinds of motion of a solid are identified: (1) translation, (2) rotation about a stationary axis, (3) plane motion, (4) motion about a stationary point, and (5) free motion. The first two kinds of motion, that is, translation and rotation about a stationary axis, are the basic kinds of motion of a solid. All the other kinds of motion of a solid prove to be reducible to one of the basic motions or to their combination. This will be shown by the example of plane motion.

In this section we shall deal with the first three kinds of motion and with the problem of summing angular velocities.

Translation. In this kind of motion of a solid any straight line fixed to it remains parallel to its initial orientation all the time. Examples: a car travelling along a straight section of a road, a Ferris wheel cage, etc.

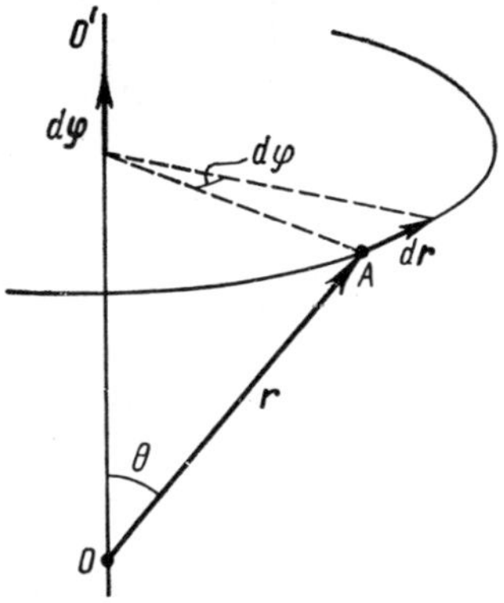

Fig. 6

When moving translationary, all points of a solid traverse equal distances in the same time interval. Therefore velocities, as well as accelerations, are of the same value at all points of the body at the given moment of time. This fact allows the study of translation of a solid to be reduced to the study of motion of an individual point belonging to that solid, i.e. to the problem of kinematics of a point.

Thus, the translation of a solid can be comprehensively described provided the dependence of the radius vector on time $\mathbf{r}(t)$ for any point of that body is available as well as the position of that body at the initial moment.

Rotation about a stationary axis. Suppose a solid, while rotating about an axis OO' which is stationary in a given reference frame, accomplishes an infinitesimal rotation during the time interval dt. We shall describe the corresponding rotation angle by the vector $d\boldsymbol{\varphi}$ whose modulus is equal to the rotation angle and whose direction coincides with the axis OO', with the rotation direction obeying the right-hand screw rule with respect to the direction of the vector $d\boldsymbol{\varphi}$ (Fig. 6).

Now let us find the elementary displacement of any point A of the solid resulting from such a rotation. The location of the point A is specified by the radius vector $\mathbf{r}$ drawn from a certain point O on the rotation axis. Then the linear displacement of the end point of the radius vector $\mathbf{r}$ is associated with the rotation angle $d\varphi$ by the relation (Fig. 6)

$$|d\mathbf{r}| = r \sin\theta \, d\varphi,$$

or in a vector form

$$d\mathbf{r} = [d\boldsymbol{\varphi}, \mathbf{r}]. \tag{1.11}$$

Note that this equality holds only for an infinitesimal rotation $d\boldsymbol{\varphi}$. In other words, only infinitesimal rotations can be treated as vectors.*

Moreover, the vector introduced ($d\boldsymbol{\varphi}$) can be shown to satisfy the basic property of vectors, that is, vector addition. Indeed, imagine a solid performing two elementary rotations, $d\boldsymbol{\varphi}_1$ and $d\boldsymbol{\varphi}_2$, about different axes crossing at a stationary point O. Then the resultant displacement $d\mathbf{r}$ of an arbitrary point A of the body, whose radius vector with respect to the point O is equal to $\mathbf{r}$, can be represented as follows:

$$d\mathbf{r} = d\mathbf{r}_1 + d\mathbf{r}_2 = [d\boldsymbol{\varphi}_1, \mathbf{r}] + [d\boldsymbol{\varphi}_2, \mathbf{r}] = [d\boldsymbol{\varphi}, \mathbf{r}],$$

where

$$d\boldsymbol{\varphi} = d\boldsymbol{\varphi}_1 + d\boldsymbol{\varphi}_2, \tag{1.12}$$

i.e. the two given rotations, $d\boldsymbol{\varphi}_1$ and $d\boldsymbol{\varphi}_2$, are equivalent to one rotation through the angle $d\boldsymbol{\varphi} = d\boldsymbol{\varphi}_1 + d\boldsymbol{\varphi}_2$ about the axis coinciding with the vector $d\boldsymbol{\varphi}$ and passing through the point O.

Note that in treating such quantities as radius vector $\mathbf{r}$, velocity $\mathbf{v}$, acceleration $\mathbf{w}$ we did not hesitate over the choice of their direction: it naturally followed from the properties of the quantities themselves. Such vectors are referred to as *polar*. As distinct from them, such vectors as $d\boldsymbol{\varphi}$ whose

---

* In the case of a finite rotation through the angle $\Delta\varphi$ the linear displacement of the point A can be found from Fig. 6:

$$|\Delta\mathbf{r}| = r \sin\theta \cdot 2 \sin(\Delta\varphi/2).$$

Whence it is immediately seen that the displacement $\Delta\mathbf{r}$ cannot be represented as a vector cross product of $\Delta\boldsymbol{\varphi}$ and $\mathbf{r}$. It is only possible in the case of an infinitesimal rotation $d\boldsymbol{\varphi}$ when the radius vector $\mathbf{r}$ can be regarded invariable.

direction is specified by the rotation direction are called *axial*.

Now let us introduce the vectors of angular velocity and angular acceleration. The angular velocity vector $\boldsymbol{\omega}$ is defined as

$$\boldsymbol{\omega} = d\boldsymbol{\varphi}/dt, \tag{1.13}$$

where dt is the time interval during which a body performs the rotation $d\boldsymbol{\varphi}$. The vector $\boldsymbol{\omega}$ is axial and its direction coincides with that of the vector $d\boldsymbol{\varphi}$.

The time variation of the vector $\boldsymbol{\omega}$ is defined by the angular acceleration vector $\boldsymbol{\beta}$:

$$\boldsymbol{\beta} = d\boldsymbol{\omega}/dt. \tag{1.14}$$

The direction of the vector $\boldsymbol{\beta}$ coincides with the direction of $d\boldsymbol{\omega}$, the increment of the vector $\boldsymbol{\omega}$. Both vectors, $\boldsymbol{\beta}$ and $\boldsymbol{\omega}$, are axial.

The representation of angular velocity and angular acceleration in a vector form proves to be very beneficial, especially in the study of more complicated kinds of motion of a solid. In many cases this makes a problem more explicit, drastically simplifies the analysis of motion and the corresponding calculations.

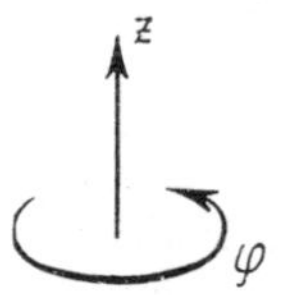

Fig. 7

Let us write the expressions for angular velocity and angular acceleration via projections on the rotation axis z whose positive direction is associated with the positive direction of the coordinate φ, the rotation angle, in accordance with the right-hand screw rule (Fig. 7). Then the projections ω_z and β_z of the vectors $\boldsymbol{\omega}$ and $\boldsymbol{\beta}$ on the z axis are defined by the following formulae:

$$\omega_z = d\varphi/dt, \tag{1.15}$$

$$\beta_z = d\omega_z/dt. \tag{1.16}$$

Here ω_z and β_z are algebraic quantities. Their sign specifies the direction of the corresponding vector. For example, if $\omega_z > 0$, then the direction of the vector $\boldsymbol{\omega}$ coincides with the positive direction of the z axis; and if $\omega_z < 0$, then the vector $\boldsymbol{\omega}$ has the opposite direction. The same is true for angular acceleration.

Thus, knowing the function $\varphi(t)$, the law of rotation of a body, we can find the angular velocity and angular acceleration at each moment of time by means of Eqs. (1.15) and (1.16) On the other hand, knowing the time dependence of angular acceleration and the initial conditions, i.e. the angular velocity ω_0 and the angle φ_0 at the initial moment of time, we can find $\omega(t)$ and $\varphi(t)$.

Example. A solid rotates about a stationary axis in accordance with the law $\varphi = at - bt^2/2$ where a and b are positive constants. Let us determine the motion characteristics of this body.

In accordance with Eqs. (1.15) and (1.16)

$$\omega_z = a - bt; \quad \beta_z = -b = \text{const}.$$

Whence it is seen that the body performs a uniformly decelerated rotation ($\beta_z < 0$), comes to a standstill at the moment $t_0 = a/b$ and then reverses its rotation direction (due to ω_z changing its sign to the opposite).

Note that the solution of all problems on rotation of a solid about a stationary axis is similar in form to that of problems on rectilinear motion of a point. It is sufficient to replace the linear quantities x, v_x and w_x by the corresponding angular quantities φ, ω_z and β_z in order to obtain all characteristics and relationships for the case of a rotating body.

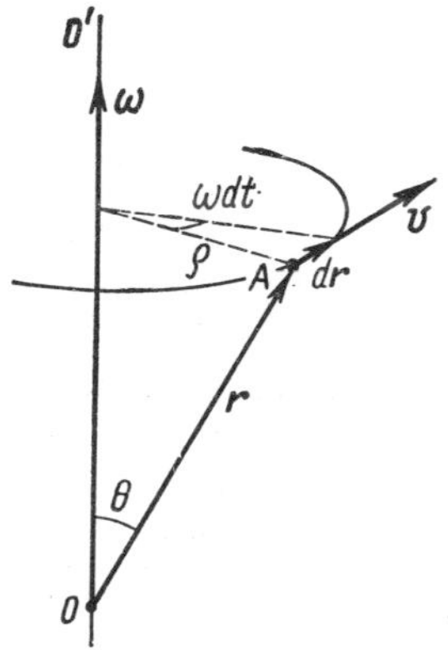

Fig. 8

Relationship between linear and angular quantities. Let us find the velocity **v** of an arbitrary point A of a solid rotating about a stationary axis OO' at an angular velocity $\boldsymbol{\omega}$. Let the location of the point A relative to some point O of the rotation axis be defined by the radius vector **r** (Fig. 8). Dividing both sides of Eq. (1.11) by the corresponding time interval dt and taking into account that $d\mathbf{r}/dt = \mathbf{v}$ and $d\boldsymbol{\varphi}/dt = \boldsymbol{\omega}$, we obtain

$$\boxed{\mathbf{v} = [\boldsymbol{\omega}\mathbf{r}],} \tag{1.17}$$

i.e. the velocity **v** of any point A of a solid rotating about some axis at an angular velocity $\boldsymbol{\omega}$ is equal to the cross product of $\boldsymbol{\omega}$ and the radius vector $\mathbf{r}$ of the point A relative to an arbitrary point O of the rotation axis (Fig. 8).

The modulus of the vector (1.17) is $v = \omega r \sin\theta$, or

$$v = \omega\rho,$$

where ρ is the radius of the circle which the point A circumscribes. Having differentiated Eq. (1.17) with respect to

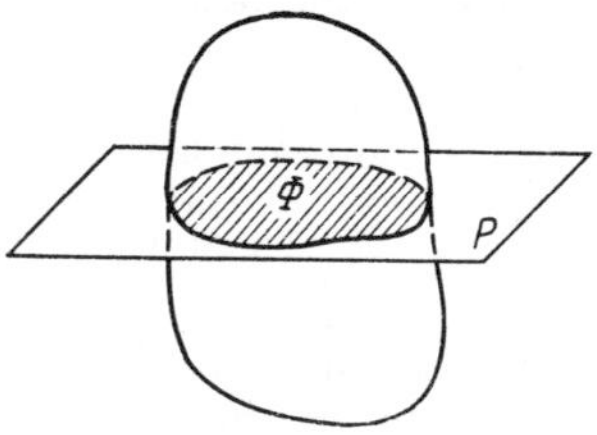

Fig. 9

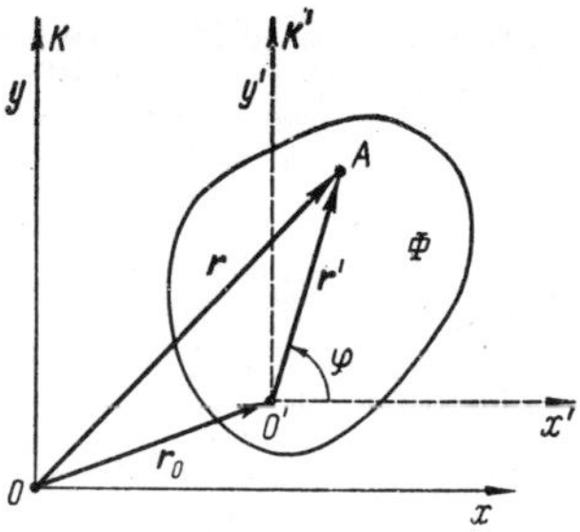

Fig. 10

time, we find the acceleration $\mathbf{w}$ of the point A:

$$\mathbf{w} = [d\boldsymbol{\omega}/dt,\ \mathbf{r}] + [\boldsymbol{\omega},\ d\mathbf{r}/dt]$$

or

$$\boxed{\mathbf{w} = [\boldsymbol{\beta}\mathbf{r}] + [\boldsymbol{\omega}\,[\boldsymbol{\omega}\mathbf{r}]].} \tag{1.18}$$

In this case (when the rotation axis is stationary) $\boldsymbol{\beta} \parallel \boldsymbol{\omega}$, and therefore the vector $[\boldsymbol{\beta}\mathbf{r}]$ represents the tangential acceleration $\mathbf{w}_\tau$. The vector $[\boldsymbol{\omega}\,[\boldsymbol{\omega}\mathbf{r}]]$ is the normal acceleration $\mathbf{w}_n$. The moduli of these vectors are

$$|\mathbf{w}_\tau| = \beta\rho; \qquad w_n = \omega^2\rho,$$

whence the modulus of the total acceleration $\mathbf{w}$ is equal to

$$w = \sqrt{w_\tau^2 + w_n^2} = \rho\sqrt{\beta^2 + \omega^4}.$$

Plane motion of a solid. In this kind of motion each point of a solid moves in a plane which is parallel to a certain stationary (in a given reference frame) plane. In this case the plane figure Φ formed as a result of cutting the solid by that stationary plane P (Fig. 9) remains in that plane all the time during the motion. Example: a cylinder rolling along a plane without slipping (in a similar case a cone performs a much more complicated motion).

It is easy to infer that the position of a solid in plane motion is unambiguously determined by the position of the plane figure Φ within the stationary plane P. The study of the plane motion of a solid thus reduces to the study of motion of a plane figure within its plane.

Let the plane figure Φ move within its plane P, which is stationary in the K reference frame (Fig. 10). The position of the figure Φ in the plane can be defined by specifying the radius vector $\mathbf{r}_0$ of an arbitrary point O' of the figure and the angle φ between the radius vector $\mathbf{r}'$ rigidly fixed to the figure and a certain selected direction in the K reference frame. The plane motion of the solid is then described by the two equations

$$\mathbf{r}_0 = \mathbf{r}_0(t); \quad \varphi = \varphi(t).$$

It is clear that if the radius vector $\mathbf{r}'$ of the point A (Fig. 10) turns through the angle $d\boldsymbol{\varphi}$ during the time interval dt, then any segment fixed to the figure will turn through the same angle. In other words, the rotation of the figure through the angle $d\boldsymbol{\varphi}$ does not depend on the choice of the point O'. This means that the angular velocity $\boldsymbol{\omega}$ of the figure does not depend on the choice of the point O', and we have the right to call $\boldsymbol{\omega}$ the angular velocity of the solid *per se*.

Now let us find the velocity $\mathbf{v}$ of an arbitrary point A of a solid in plane motion. Let us introduce the auxiliary reference frame K' which is rigidly fixed to the point O' of the solid and translates relative to the K frame (Fig. 10). Then the elementary displacement $d\mathbf{r}$ of the point A in the K frame can be written in the following form:

$$d\mathbf{r} = d\mathbf{r}_0 + d\mathbf{r}',$$

where $d\mathbf{r}_0$ is the displacement of the K' frame, or the point O', and $d\mathbf{r}'$ is the displacement of the point A relative to the K' frame. The translation $d\mathbf{r}'$ is caused by the rotation of the solid about the axis which is at rest in the K' frame and passes through the point O'; according to Eq. (1.11) $d\mathbf{r}' = [d\boldsymbol{\varphi}, \mathbf{r}']$. Substituting this relation into the previous one and dividing both sides of the expression obtained by dt, we get

$$\mathbf{v} = \mathbf{v}_0 + [\boldsymbol{\omega}\mathbf{r}'], \tag{1.19}$$

i.e. the velocity of any point A of a solid in plane motion* comprises the velocity $\mathbf{v}_0$ of an arbitrary point O' of that solid and the velocity $\mathbf{v}' = [\boldsymbol{\omega}\mathbf{r}']$ caused by rotation of the solid about the axis passing through the point O'. Once again we would like to stress that $\mathbf{v}'$ is the velocity of the point A relative to the translating reference frame K' which is rigidly fixed to the point O'.

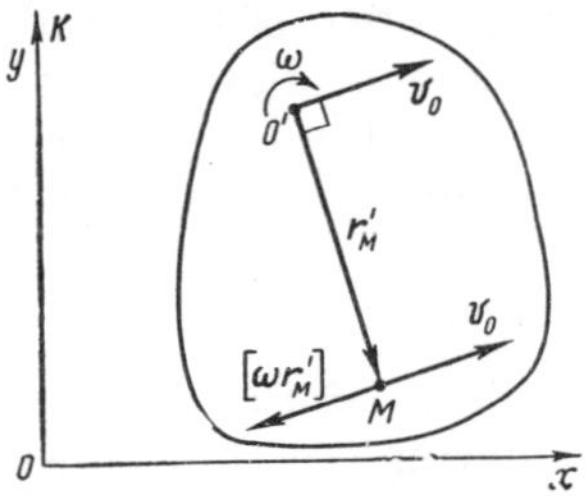

Fig. 11

In other words, plane motion of a solid can be represented as a combination of two basic kinds of motion: translation (together with an arbitrary point O' of the solid) and rotation (around an axis passing through the point O').

Now we shall demonstrate that plane motion can be reduced to a purely rotational motion. Indeed, in plane motion the velocity $\mathbf{v}_0$ of the arbitrary point O' of the solid is normal to the vector $\boldsymbol{\omega}$ which means that we can always find a certain point M which is rigidly fixed to the solid** and whose velocity $\mathbf{v} = 0$ at a given moment. The location of the point M, i.e. its radius vector $\mathbf{r}'_M$ relative to the point O' (Fig. 11), can be found from the condition $0 = \mathbf{v}_0 + [\boldsymbol{\omega}\mathbf{r}'_M]$. The vector $\mathbf{r}'_M$ is perpendicular to $\boldsymbol{\omega}$ and $\mathbf{v}_0$, its direction corresponding to the vector cross product $\mathbf{v}_0 = -[\boldsymbol{\omega}\mathbf{r}'_M]$ and its magnitude $r'_M = v_0/\omega$.

The point M defines the position of another important axis (coinciding with the direction of the vector $\boldsymbol{\omega}$). At a given moment of time the motion of a solid represents a pure rotation about this axis. Such an axis is referred to as an *instantaneous rotation axis*.

---

* Note that Eq. (1.19) also holds for any complex motion of a solid.

** The point M may turn out to be outside the solid.

Generally speaking, the position of the instantaneous axis varies with time. For example, in the case of a cylinder rolling over a plane surface the instantaneous axis coincides at any moment with the line of contact between the cylinder and the plane.

Angular velocity summation. Let us analyse the motion of a solid rotating simultaneously about two intersecting axes. We shall set into rotation a certain solid at the angular velocity $\boldsymbol{\omega}'$ about the axis *OA* (Fig. 12), and then we shall set this axis into rotation with the angular velocity $\boldsymbol{\omega}_0$ about the axis *OB* which is stationary in the *K* reference frame. Let us find the resultant motion in the *K* frame.

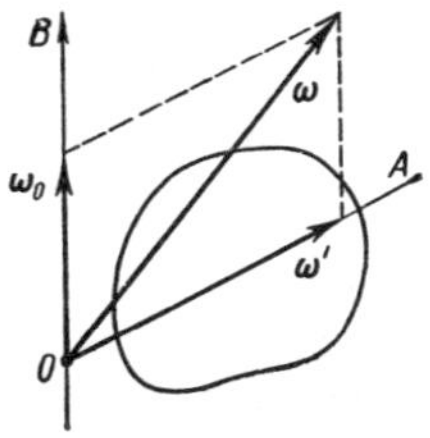

Fig. 12

We shall introduce an auxiliary reference frame *K'* fixed rigidly to the axes *OA* and *OB*. It is clear that this frame rotates with the angular velocity $\boldsymbol{\omega}_0$ while the solid rotates relative to this frame with the angular velocity $\boldsymbol{\omega}'$.

During the time interval dt the solid will turn through an angle $d\boldsymbol{\varphi}'$ about the axis *OA* in the *K'* frame and simultaneously through $d\boldsymbol{\varphi}_0$ about the axis *OB* together with the *K'* frame. The cumulative rotation follows from Eq. (1.12): $d\boldsymbol{\varphi} = d\boldsymbol{\varphi}_0 + d\boldsymbol{\varphi}'$. Dividing both sides of this equality by dt, we obtain

$$\boldsymbol{\omega} = \boldsymbol{\omega}_0 + \boldsymbol{\omega}'. \tag{1.20}$$

Thus, the resultant motion of the solid in the *K* frame is a pure rotation with the angular velocity $\boldsymbol{\omega}$ about an axis coinciding at each moment with the vector $\boldsymbol{\omega}$ and passing through the point *O* (Fig. 12). This axis is displaced relative to the *K* frame: it rotates together with the *OA* axis about the axis *OB* at the angular velocity $\boldsymbol{\omega}_0$.

It is not difficult to infer that even when the angular velocities $\boldsymbol{\omega}'$ and $\boldsymbol{\omega}_0$ do not change their magnitudes, the body in the *K* frame will possess the angular acceleration $\boldsymbol{\beta}$ directed, according to Eq. (1.14), beyond the plane (Fig. 12). The angular acceleration of a solid is analysed in detail in Problem 1.10.

And here is one more remark. Since the angular velocity vector $\boldsymbol{\omega}$ satisfies the basic property of vectors, vector summation, $\boldsymbol{\omega}$ can be expanded into a sum of vector components projected on definite directions, i.e. $\boldsymbol{\omega} = \boldsymbol{\omega}_1 + \boldsymbol{\omega}_2 + \ldots$, where all vectors belong to the same reference frame. This convenient and beneficial routine is frequently employed to analyse complex motions of a solid.

§ 1.3. Transformation of Velocity and Acceleration on Transition to Another Reference Frame

Prior to entering upon the study of this problem we should recall that within the bounds of classical mechanics the length of scales and time are considered absolute. A scale is the same in different reference frames, i.e. it does not change during motion. This is also true of time running uniformly throughout all frames.

Formulation of the problem. There are two arbitrary reference frames K and K' moving relative to each other in a definite manner. The velocity $\mathbf{v}$ and the acceleration $\mathbf{w}$ of a point A in the K frame are known. What are the corresponding values $\mathbf{v}'$ and $\mathbf{w}'$ of this point in the K' frame?

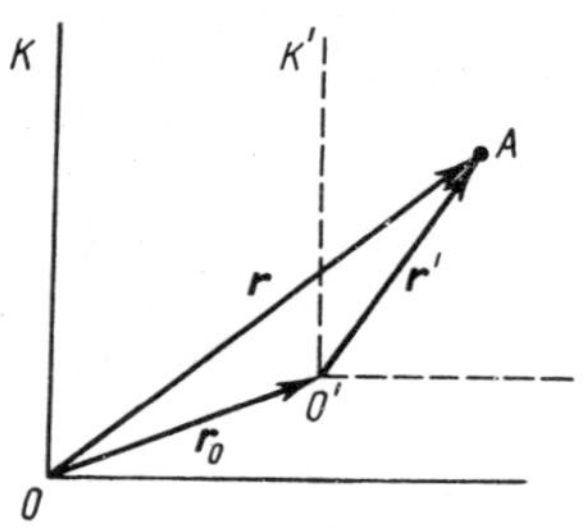

Fig. 13

We shall examine the three most significant cases of relative motion of two reference frames in succession.

1. ***The K' frame translates relative to the K frame.*** Suppose the origin of the K' frame is determined by the radius vector $\mathbf{r}_0$ in the K frame, and its velocity and acceleration by the vectors $\mathbf{v}_0$ and $\mathbf{w}_0$. If the location of the point A in the K frame is determined by the radius vector $\mathbf{r}$ and in the K' frame by the radius vector $\mathbf{r}'$, then apparently $\mathbf{r} = \mathbf{r}_0 + \mathbf{r}'$ (Fig. 13). Next, let during the time interval dt the point A accomplish the elementary displacement $d\mathbf{r}$ in the K frame. This displacement is made up of the displacement $d\mathbf{r}_0$ (together with the K' frame) and the displace-

ment $d\mathbf{r}'$ relative to the K' frame: $d\mathbf{r} = d\mathbf{r}_0 + d\mathbf{r}'$. Dividing this expression by dt, we obtain the following formula for the velocity transformation:

$$\boxed{\mathbf{v} = \mathbf{v}_0 + \mathbf{v}'.} \tag{1.21}$$

Differentiating Eq. (1.21) with respect to time, we immediately get the acceleration transformation formula:

$$\boxed{\mathbf{w} = \mathbf{w}_0 + \mathbf{w}'.} \tag{1.22}$$

Whence it is seen, specifically, that if $\mathbf{w}_0 = 0$ and $\mathbf{w} = \mathbf{w}'$, i.e. when the K' frame moves without acceleration, the acceleration of the point A relative to the K frame will be the same in both frames.

2. ***The K' frame rotates at the constant angular velocity ω about an axis which is stationary in the K frame.***

Let us assume the origins of the reference frames K and K' to be located at an arbitrary point O on the rotation axis

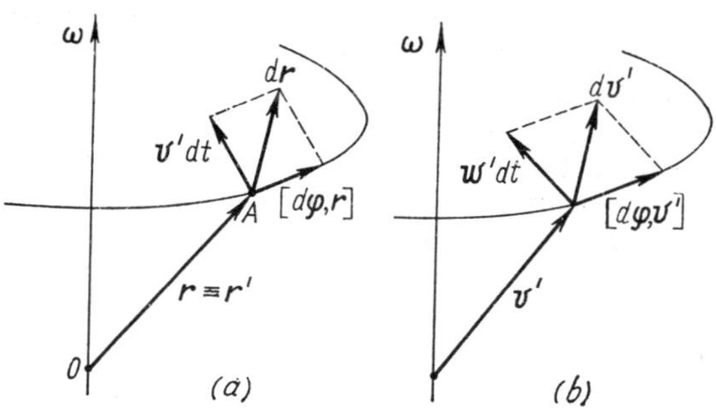

Fig. 14

(Fig. 14a). Then the radius vector of the point A will be the same in both reference frames: $\mathbf{r} \equiv \mathbf{r}'$.

If the point A is at rest in the K' frame, this means that its displacement $d\mathbf{r}$ in the K frame during the time interval dt is caused only by the rotation of the radius vector $\mathbf{r}$ through the angle $d\varphi$ (together with the K' frame) and in accordance with Eq. (1.11) is equal to the vector cross product $[d\boldsymbol{\varphi}, \mathbf{r}]$.

If the point A moves at the velocity $\mathbf{v}'$ relative to the K' frame, it will cover an additional distance $\mathbf{v}'dt$ during the time interval dt (Fig. 14a), so that

$$d\mathbf{r} = \mathbf{v}'dt + [d\boldsymbol{\varphi}, \mathbf{r}]. \tag{1.23}$$

Dividing this expression by dt, we obtain the velocity transformation formula as follows:

$$\boxed{\mathbf{v} = \mathbf{v}' + [\boldsymbol{\omega}\mathbf{r}],} \tag{1.24}$$

where $\mathbf{v}$ and $\mathbf{v}'$ are the velocity values which characterize the motion of the point A in the K and K' frames respectively.

Now let us pass over to acceleration. In accordance with Eq. (1.24) the increment $d\mathbf{v}$ of the vector $\mathbf{v}$ during the time interval dt in the K frame must comprise the sum of the increments of the vectors $\mathbf{v}'$ and $[\boldsymbol{\omega}\mathbf{r}]$, i.e.

$$d\mathbf{v} = d\mathbf{v}' + [\boldsymbol{\omega}, d\mathbf{r}]. \tag{1.25}$$

Let us find $d\mathbf{v}'$. If the point A moves in the K' frame with a constant velocity ($\mathbf{v}'$ = const), the increment of this vector in the K frame is caused only by this vector turning through the angle $d\boldsymbol{\varphi}$ (together with the K' frame) and is equal, as in the case of $\mathbf{r}$, to the vector cross product $[d\boldsymbol{\varphi}, \mathbf{v}']$. To make sure of this, let us position the beginning of the vector $\mathbf{v}'$ on the rotation axis (Fig. 14b). But if the point A moves with the acceleration $\mathbf{w}'$ in the K' frame, the vector $\mathbf{v}'$ will get an additional increment $\mathbf{w}'dt$ during the time interval dt, and consequently

$$d\mathbf{v}' = \mathbf{w}'dt + [d\boldsymbol{\varphi}, \mathbf{v}']. \tag{1.26}$$

Now let us substitute Eqs. (1.26) and (1.23) into Eq. (1.25) and then divide the expression obtained by dt. Thus we shall get the acceleration transformation formula:

$$\mathbf{w} = \mathbf{w}' + 2[\boldsymbol{\omega}\mathbf{v}'] + [\boldsymbol{\omega}[\boldsymbol{\omega}\mathbf{r}]], \tag{1.27}$$

where $\mathbf{w}$ and $\mathbf{w}'$ are the acceleration values of the point A observed in the K and K' frames. The second term on the right-hand side of this formula is referred to as the *Coriolis* acceleration $\mathbf{w}_{Cor}$ and the third term is the *axipetal* acceleration $\mathbf{w}_{ap}$ directed toward the axis*

$$\mathbf{w}_{Cor}=2[\boldsymbol{\omega}\mathbf{v}'], \qquad \mathbf{w}_{ap}=[\boldsymbol{\omega}[\boldsymbol{\omega}\mathbf{r}]]. \tag{1.28}$$

Thus, the acceleration $\mathbf{w}$ of the point relative to the K frame is equal to the sum of three accelerations: the acceleration $\mathbf{w}'$ relative to the K' frame, the Coriolis acceleration $\mathbf{w}_{Cor}$ and the axipetal acceleration $\mathbf{w}_{ap}$.

The axipetal acceleration can be represented in the form $\mathbf{w}_{ap} = -\omega^2\boldsymbol{\rho}$ where $\boldsymbol{\rho}$ is the radius vector which is normal to the rotation axis and describes the position of the point A relative to this axis. Then Eq. (1.27) can be written as follows:

$$\boxed{\mathbf{w}=\mathbf{w}'+2[\boldsymbol{\omega}\mathbf{v}']-\omega^2\boldsymbol{\rho}.} \tag{1.29}$$

3. ***The K′ frame rotates with the constant angular velocity $\boldsymbol{\omega}$ about the axis translating with the velocity $\mathbf{v}_0$ and acceleration $\mathbf{w}_0$ relative to the K frame.***

This case combines the two previous ones. Let us introduce an auxiliary S frame which is rigidly fixed to the rotation axis of the K' frame and translates in the K frame. Suppose $\mathbf{v}$ and $\mathbf{v}_S$ are the velocity values of the point A in the K and S frames; then in accordance with Eq. (1.21) $\mathbf{v} = \mathbf{v}_0 + \mathbf{v}_S$. Replacing $\mathbf{v}_S$ in accordance with Eq. (1.24) by $\mathbf{v}_S = \mathbf{v}' + [\boldsymbol{\omega}\mathbf{r}]$, where $\mathbf{r}$ is the radius vector of the point A relative to the arbitrary point on the rotation axis of the K' frame, we obtain the following velocity transformation formula:

$$\boxed{\mathbf{v}=\mathbf{v}'+\mathbf{v}_0+[\boldsymbol{\omega}\mathbf{r}].} \tag{1.30}$$

* This axipetal acceleration should not be confused with conventional (centripetal) acceleration.

In a similar fashion, using Eqs. (1.22) and (1.29), we obtain the acceleration transformation formula*:

$$\boxed{\mathbf{w}=\mathbf{w}'+\mathbf{w}_0+2[\boldsymbol{\omega}\mathbf{v}']-\omega^2\boldsymbol{\rho}.} \tag{1.31}$$

Recall that in the last two formulae $\mathbf{v}$, $\mathbf{v}'$ and $\mathbf{w}$, $\mathbf{w}'$ are the velocities and accelerations of the point A in the K and K' frames respectively, $\mathbf{v}_0$ and $\mathbf{w}_0$ are the velocity and acceleration of the rotation axis of the K' frame in the K frame, $\mathbf{r}$ is the radius vector of the point A relative to an arbitrary point on the rotation axis of the K' frame, and $\boldsymbol{\rho}$ is the radius vector perpendicular to the rotation axis and describing the location of the point A relative to this axis.

In conclusion, let us examine the following example.

Example. A disc rotates with a constant angular velocity $\boldsymbol{\omega}$ about an axis fixed to the table. Point A moves along the disc with the constant velocity $\mathbf{v}$ relative to the table. Find the velocity $\mathbf{v}'$ and acceleration $\mathbf{w}'$ of the point A relative to the disc at the moment when the radius vector describing its position relative to the rotation axis is equal to $\boldsymbol{\rho}$.

In accordance with Eq. (1.24) the velocity $\mathbf{v}'$ of the point A is equal to

$$\mathbf{v}'=\mathbf{v}-[\boldsymbol{\omega}\boldsymbol{\rho}].$$

The acceleration $\mathbf{w}'$ can be found from Eq. (1.29), taking into account that in this case $\mathbf{w}=0$ since $\mathbf{v}=\text{const}$. Then $\mathbf{w}'=-2[\boldsymbol{\omega}\mathbf{v}']+$ $+\omega^2\boldsymbol{\rho}$. Substituting the expression for $\mathbf{v}'$ into this formula we obtain

$$\mathbf{w}'=2[\mathbf{v}\boldsymbol{\omega}]-\omega^2\boldsymbol{\rho}.$$

Problems to Chapter 1

●**1.1.** The radius vector describing the position of the particle A relative to the stationary point O changes with time according to the following law:

$$\mathbf{r}=\mathbf{a}\sin\omega t+\mathbf{b}\cos\omega t,$$

where $\mathbf{a}$ and $\mathbf{b}$ are constant vectors, with $\mathbf{a}\perp\mathbf{b}$; ω is a positive constant. Find the acceleration $\mathbf{w}$ of the particle and the equation of its path $y(x)$, assuming the x and y axes to coincide with the directions of the vectors $\mathbf{a}$ and $\mathbf{b}$ respectively and to have the origin at the point O.

---

* Note that in the most general case when $\omega\neq\text{const}$, the right-hand side of Eq. (1.31) will feature one more term, namely $[\boldsymbol{\beta}\mathbf{r}]$, where $\boldsymbol{\beta}$ is the angular acceleration of the K' frame, $\mathbf{r}$ is the radius vector describing the position of the point located on the rotation axis and taken for the origin in the K' frame.

Solution. Differentiating **r** with respect to time twice, we obtain

$$\mathbf{w} = -\omega^2 (\mathbf{a} \sin \omega t + \mathbf{b} \cos \omega t) = -\omega^2 \mathbf{r},$$

i.e. the vector **w** is always oriented toward the point O while its magnitude is proportional to the distance between the particle and the point O.

Now let us determine the trajectory equation. Projecting **r** on the x and y axes, we obtain

$$x = a \sin \omega t, \qquad y = b \cos \omega t.$$

Eliminating ωt from these two equations, we get

$$x^2/a^2 + y^2/b^2 = 1.$$

This is the equation of an ellipse, and a and b are its semi-axes (see Fig. 15; the arrow shows the direction of motion of the particle A).

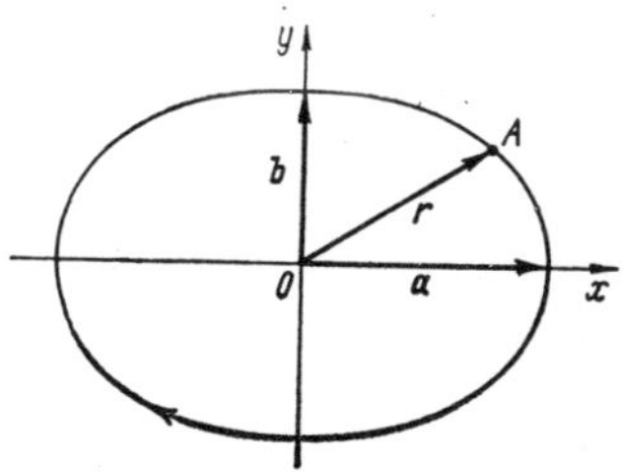

Fig. 15

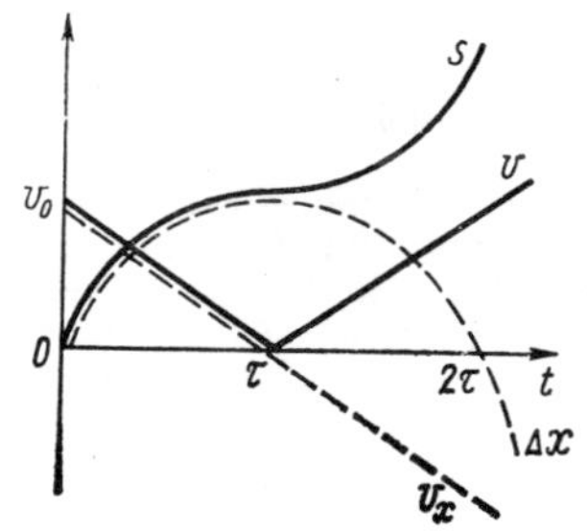

Fig. 16

1.2. Displacement and distance. At the moment $t = 0$ a particle is set in motion at the velocity $\mathbf{v}_0$ whereupon its velocity begins changing with time in accordance with the law

$$\mathbf{v} = \mathbf{v}_0 (1 - t/\tau),$$

where τ is a positive constant. Find:

(1) the displacement vector $\Delta\mathbf{r}$ of the particle, and

(2) the distance s covered by it in the first t seconds of motion.

Solution. 1. In accordance with Eq. (1.1) $d\mathbf{r} = \mathbf{v}\,dt = \mathbf{v}_0 (1 - t/\tau)\,dt$. Integrating this equation with respect to time between O and t, we obtain

$$\Delta\mathbf{r} = \mathbf{v}_0 t (1 - t/2\tau).$$

2. The distance s covered by the particle in the time t is determined by

$$s = \int_0^t v\,dt,$$

where v is the modulus of the vector **v**. In this case

$$v=v_0|1-t/\tau|=\begin{cases} v_0(1-t/\tau), & \text{if } t\leqslant\tau, \\ v_0(t/\tau-1), & \text{if } t\geqslant\tau. \end{cases}$$

From this it follows that if $t>\tau$ the integral for calculating the distance should be subdivided into two parts: between 0 and τ and between τ and t. Integrating in the two cases ($t<\tau$ and $t>\tau$), we obtain

$$s=\begin{cases} v_0t(1-t/2\tau), & \text{if } t\leqslant\tau, \\ v_0\tau[1+(1-t/\tau)^2]/2, & \text{if } t\geqslant\tau. \end{cases}$$

Fig. 16 illustrates the plots $v(t)$ and $s(t)$. The dotted lines show the time dependences of the projections v_x and Δx of the vectors **v** and Δ**r** on the x axis oriented along the vector $\mathbf{v}_0$.

●**1.3.** A street car moves rectilinearly from station A to the next stop B with an acceleration varying according to the law $w=a-bx$ where a and b are positive constants and x is its distance from station A. Find the distance between these stations and the maximum velocity of the street car.

Solution. First we shall find how the velocity depends on x. During the time interval dt the velocity increment $dv=w\,dt$. Making use of the equation $dt=dx/v$, we reduce the last expression to the form which is convenient to integrate:

$$v\,dv=(a-bx)\,dx.$$

Integrating this equation (the left-hand side between 0 and v and the right-hand side between 0 and x), we get

$$v^2/2=ax-bx^2/2 \quad \text{or} \quad v=\sqrt{(2a-bx)\,x}.$$

From this equation it can be immediately seen that the distance between the stations, that is, the value x_0 corresponding to $v=0$ is equal to $x_0=2a/b$. The maximum velocity can be found from the condition $dv/dx=0$, or, simply, from the condition for the maximum value of the radicand. The value x_m corresponding to v_{max} is equal to $x_m=a/b$ and $v_{max}=a/\sqrt{b}$.

●**1.4.** A particle moves in the x, y plane from the point $x=y=0$ with the velocity $\mathbf{v}=a\mathbf{i}+bx\mathbf{j}$, where a and b are constants and **i** and **j** are the unit vectors of the x and y axes. Find the equation of its path $y(x)$.

Solution. Let us write the increments of the x and y coordinates of the particle in the time interval dt: $dy=v_y\,dt$, $dx=v_x\,dt$, where $v_y=bx$, $v_x=a$. Taking their ratio, we get

$$dy=(b/a)\,x\,dx.$$

Integrating this expression, we obtain the following equation:

$$y=\int_0^x (b/a)\, x\, dx=(b/2a)\, x^2,$$

i.e. the path of the point is a parabola.

● **1.5.** The motion law for the point A of the rim of a wheel rolling uniformly along a horizontal path (the x axis) has the form

$$x = a\,(\omega t - \sin \omega t); \qquad y = a\,(1 - \cos \omega t),$$

where a and ω are positive constants. Find the velocity v of the point A, the distance s which it traverses between two successive contacts with the roadbed, as well as the magnitude and the direction of the acceleration $\mathbf{w}$ of the point A.

Solution. The velocity v of the point A and the distance s it covers are determined by the formulae

$$v=\sqrt{v_x^2+v_y^2}=a\omega\sqrt{2\,(1-\cos \omega t)}=2a\omega \sin (\omega t/2),$$

$$s=\int_0^{t_1} v\,(t)\, dt=4a\,[1-\cos (\omega t_1/2)],$$

where t_1 is the time interval between two successive contacts. From $y\,(t)$ we find that $y\,(t_1) = 0$ at $\omega t_1 = 2\pi$. Therefore, $s = 8a$.

The acceleration of the point A

$$w=\sqrt{w_x^2+w_y^2}=a\omega^2.$$

Let us show that the vector $\mathbf{w}$, constant in its magnitude, is always directed toward the centre of the wheel, the point C. In fact, in the K' frame fixed to the point C and translating uniformly relative to the roadbed the point A moves uniformly along a circle about the point C. Consequently, its acceleration in the K' frame is directed toward the centre of the wheel. And since the K' frame moves uniformly, the vector $\mathbf{w}$ is the same relative to the roadbed.

● **1.6.** A point moves along a circle of radius r with deceleration; at any moment the magnitudes of its tangential and normal accelerations are equal. The point was set in motion with the velocity v_0. Find the velocity v and the magnitude of the total acceleration w of the point as a function of the distance s covered by it.

Solution. By the hypothesis, $dv/dt = -v^2/r$. Replacing dt by ds/v, we reduce the initial equation to the form

$$dv/v = -ds/r.$$

The integration of this expression with regard to the initial velocity yields the following result:

$$v=v_0 e^{-s/r}.$$

In this case $|w_\tau| \equiv w_n$, and therefore the total acceleration $w = \sqrt{2}\, w_n = \sqrt{2}\, v^2/r$, or

$$w = \sqrt{2}\, v_0^2/re^{2s/r}.$$

●1.7. A point moves along a plane path so that its tangential acceleration $w_\tau = a$ and the normal acceleration $w_n = bt^4$, where a and b are positive constants and t is time. The point started moving at the moment $t = 0$. Find the curvature radius ρ of its path and its total acceleration w as a function of the distance s covered by the point.

Solution. The elementary velocity increment of the point $dv = w_\tau\, dt$. Integrating this equation, we get $v = at$. The distance covered $s = at^2/2$.

In accordance with Eq. (1.10) the curvature radius of the path can be represented as $\rho = v^2/w_n = a^2/bt^2$, or

$$\rho = a^3/2bs.$$

The total acceleration

$$w = \sqrt{w_\tau^2 + w_n^2} = a\sqrt{1 + (4bs^2/a^3)^2}.$$

●1.8. A particle moves uniformly with the velocity v along a parabolic path $y = ax^2$, where a is a positive constant. Find the acceleration w of the particle at the point $x = 0$.

Solution. Let us differentiate twice the path equation with respect to time:

$$\frac{dy}{dt} = 2ax\frac{dx}{dt}; \qquad \frac{d^2y}{dt^2} = 2a\left[\left(\frac{dx}{dt}\right)^2 + x\frac{d^2x}{dt^2}\right].$$

Since the particle moves uniformly, its acceleration at all points of the path is purely normal and at the point $x = 0$ it coincides with the derivative d^2y/dt^2 at that point. Keeping in mind that at the point $x = 0$ $|dx/dt| = v$, we get

$$w = (d^2y/dt^2)_{x=0} = 2av^2.$$

Note that in this solution method we have avoided calculating the curvature radius of the path at the point $x = 0$, which is usually needed to determine the normal acceleration ($w_n = v^2/\rho$).

●1.9. **Rotation of a solid.** A solid starts rotating about a stationary axis with the angular acceleration $\boldsymbol{\beta} = \boldsymbol{\beta}_0 \cos\varphi$, where $\boldsymbol{\beta}_0$ is a constant vector and φ is the angle of rotation of the solid from the initial position. Find the angular velocity ω_z of the solid as a function of φ.

Solution. Let us choose the positive direction of the z axis along the vector $\boldsymbol{\beta}_0$. In accordance with Eq. (1.16) $d\omega_z = \beta_z\, dt$. Using Eq. (1.15) to replace dt by $d\varphi/\omega_z$, we reduce the previous equation to the following form:

$$\omega_z\, d\omega_z = \beta_0 \cos\varphi\, d\varphi.$$

The integration of this expression with regard to the initial condition ($\omega_z = 0$ at $\varphi = 0$) yields $\omega_z^2/2 = \beta_0 \sin \varphi$. From this it follows that

$$\omega_z = \pm \sqrt{2\beta_0 \sin \varphi}.$$

The plot $\omega_n(\varphi)$ is shown in Fig. 17. It can be seen that as the angle φ grows, the vector $\boldsymbol{\omega}$ first increases, coinciding with the direction of the vector $\boldsymbol{\beta}_0$ ($\omega_z > 0$), reaches the maximum at $\varphi = \pi/2$, then starts decreasing and finally turns into zero at $\varphi = \pi$. After that the body starts rotating in the opposite direction in a similar fashion ($\omega_z < 0$). As a result, the body will oscillate about the position $\varphi = \pi/2$ with an amplitude equal to $\pi/2$.

●1.10. A round cone having the height h and the base radius r rolls without slipping along the table surface as shown in Fig. 18.

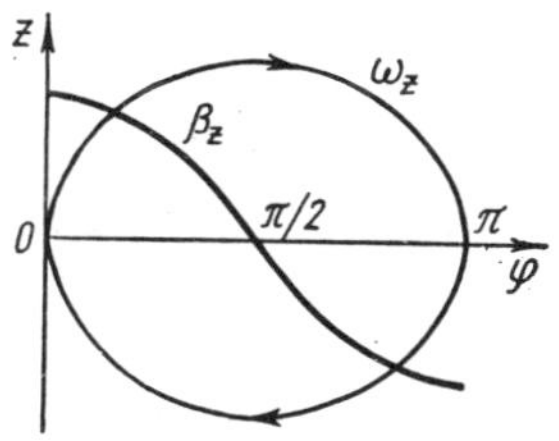

Fig. 17

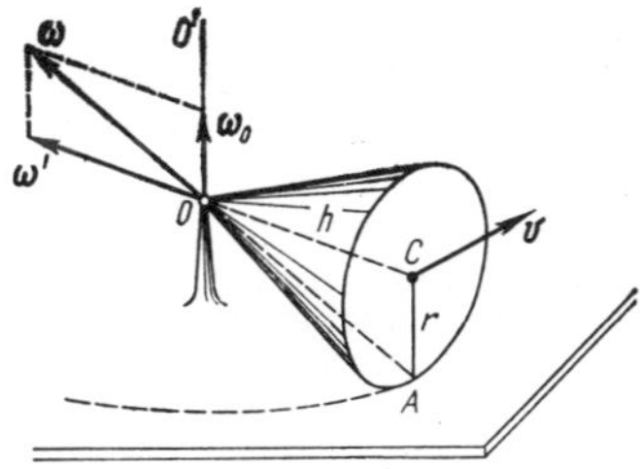

Fig. 18

The cone apex is hinged at the point O which is exactly level with the point C, the cone base centre. The point C moves at the constant velocity v. Find:

(1) the angular velocity $\boldsymbol{\omega}$ and

(2) the angular acceleration $\boldsymbol{\beta}$ of the cone relative to the table.

Solution. 1. In accordance with Eq. (1.20) $\boldsymbol{\omega} = \boldsymbol{\omega}_0 + \boldsymbol{\omega}'$, where $\boldsymbol{\omega}_0$ and $\boldsymbol{\omega}'$ are the angular velocities of rotation about the axes OO' and OC respectively. The magnitudes of the vectors $\boldsymbol{\omega}_0$ and $\boldsymbol{\omega}'$ can be easily found from Fig. 18:

$$\omega_0 = v/h, \quad \omega' = v/r.$$

Their ratio $\omega_0/\omega' = r/h$. It follows that the vector $\boldsymbol{\omega}$ coincides at any moment with the cone generatrix which passes through the contact point A.

The magnitude of the vector $\boldsymbol{\omega}$ is equal to

$$\omega = \sqrt{\omega_0^2 + \omega'^2} = (v/r)\sqrt{1 + (r/h)^2}.$$

2. In accordance with Eq. (1.14) the angular acceleration $\boldsymbol{\beta}$ of the cone is represented by the derivative of the vector $\boldsymbol{\omega}$ with respect to time. Since $\boldsymbol{\omega}_0 = \text{const}$, then

$$\boldsymbol{\beta} = d\boldsymbol{\omega}/dt = d\boldsymbol{\omega}'/dt.$$

The vector $\boldsymbol{\omega}'$ rotating about the OO' axis with the angular velocity ω_0 retains its magnitude. Its increment in the time interval dt is equal to $|d\boldsymbol{\omega}'| = \omega' \cdot \omega_0 \, dt$, or in vector form to $d\boldsymbol{\omega}' = [\boldsymbol{\omega}_0 \boldsymbol{\omega}'] \, dt$. Thus,

$$\boldsymbol{\beta} = [\boldsymbol{\omega}_0 \boldsymbol{\omega}'].$$

The magnitude of this vector $\boldsymbol{\beta}$ is equal to $\beta = v^2/rh$.

● **1.11. Velocity and acceleration transformation.** A horizontal bar rotates with the constant angular velocity $\boldsymbol{\omega}$ about a vertical axis which is fixed to a table and passes through one of the ends of that bar. A small coupling moves along the bar. Its velocity relative to the bar obeys the law $\mathbf{v}' = a\mathbf{r}$ where a is a constant and $\mathbf{r}$ is the radius vector determining the distance between the coupling and the rotation axis. Find:

(1) the velocity $\mathbf{v}$ and the acceleration $\mathbf{w}$ of the coupling relative to the table and depending on $\mathbf{r}$;

(2) the angle between the vectors $\mathbf{v}$ and $\mathbf{w}$ in the process of motion.

Solution. 1. In accordance with Eq. (1.24)

$$\mathbf{v} = a\mathbf{r} + [\boldsymbol{\omega}\mathbf{r}].$$

The magnitude of this vector $v = r\sqrt{a^2 + \omega^2}$.

The acceleration $\mathbf{w}$ is found from Eq. (1.29) where in this case $\mathbf{w}' = d\mathbf{v}'/dt = a^2\mathbf{r}$. Then

$$\mathbf{w} = (a^2 - \omega^2)\,\mathbf{r} + 2a\,[\boldsymbol{\omega}\mathbf{r}].$$

The magnitude of this vector $w = (a^2 + \omega^2)\,r$.

2. To calculate the angle α between the vectors $\mathbf{v}$ and $\mathbf{w}$, we shall make use of their scalar product, from which it follows that $\cos\alpha = \mathbf{v}\mathbf{w}/vw$. After the requisite transformations we obtain

$$\cos\alpha = 1/\sqrt{1 + (\omega/a)^2}.$$

It is seen from this formula that in this case the angle α remains constant during the motion.

CHAPTER 2

THE BASIC EQUATION OF DYNAMICS

§ 2.1. Inertial Reference Frames

The law of inertia. Kinematics, being concerned with describing motion irrespective of its causes, makes no essential difference between various reference frames and regards them as equivalent. It is quite different with dynamics, which deals with laws of motion. Here we detect the intrinsic difference between various reference frames and identify the advantages of one class of frames over others.

Basically, we can use any one of the infinite number of reference frames. But the laws of mechanics have, generally speaking, a different form in different reference frames; it may then happen that in an arbitrary reference frame the laws governing simple phenomena prove to be very complicated. Thus, we face the problem of choosing a reference frame in which the laws of mechanics take the simplest form. Such a reference frame is obviously most suitable for describing mechanical phenomena.

With this aim in view let us consider acceleration of a mass point relative to an arbitrary reference frame. What causes the acceleration? Experience shows that it can be due to some definite bodies acting on this point, as well as to the properties of the reference frame itsels (in fact, in the general case the acceleration is different relative to different reference frames).

We can, however, *assume* that there is a reference frame in which acceleration of a mass point arises solely due to its interaction with other bodies. Then a free mass point experiencing no action from any other bodies moves rectilinearly and uniformly, relative to such a frame, or, in other words, due to inertia. Such a reference frame is called *inertial*.

The statement of the existence of inertial reference frames formulates the content of the **first law of classical mechanics**, *the law of inertia of Galileo and Newton.*

The existence of inertial frames is corroborated by experiments. By early tests it was established that the Earth represents such a frame. Subsequently, the more accurate experiments (Foucault's experiment and the like) argued that this reference frame is not totally inertial*, viz., some kinds of acceleration were detected whose occurrence cannot be explained by any definite bodies acting in this frame. At the same time the observation of acceleration of planets proved the inertial character of the heliocentric reference frame fixed to the centre of the sun and "stationary" stars. At the present time the inertial character of the heliocentric reference frame is confirmed by the whole totality of experimental facts.

Any other reference frame moving rectilinearly and uniformly relative to the heliocentric frame is also inertial. In fact, if the acceleration of a body is equal to zero in the heliocentric reference frame, it will be equal to zero in any other of these reference frames.

Thus, there is a vast number of inertial reference frames moving relative to one another rectilinearly and uniformly. Reference frames executing accelerated motion relative to inertial ones are called *non-inertial*.

On symmetry properties of time and space. An important feature of inertial frames consists in the fact that time and space possess definite *symmetry properties* with respect to them. Specifically, experience shows that in such frames time is *uniform* while space is both *uniform and isotropic*.

The uniformity of time signifies that physical phenomena proceed identically at different moments when observed under the same conditions. In other words, different moments of time are equivalent in terms of their physical properties.

The uniformity and isotropy of space mean that the properties of space are identical at all points (uniformity) and in all directions at each point (isotropy).

Note that space is non-uniform and anisotropic with respect to non-inertial reference frames. This means that if a certain body does not interact with any other bodies, its different orientations are still not equivalent in mechanical terms. In the general case this is also true for time which is non-uniform, i.e. different moments of time are not equiva-

---

* It should be pointed out that in many cases the reference frame fixed to the Earth can be regarded practically inertial.

lent. It is clear that such properties of space and time would complicate the description of mechanical phenomena very much. For example, a body experiencing no action from other bodies could not be at rest: even though its velocity is equal to zero at an initial moment of time, the next moment the body would start moving in a definite direction.

Galilean relativity. In inertial reference frames the following principle of relativity is valid: all inertial frames are equivalent in their mechanical properties. This means that no mechanical tests performed "inside" a given inertial frame can detect whether that frame moves or not. Throughout all inertial reference frames the properties of space and time, as well as all laws of mechanics, are identical.

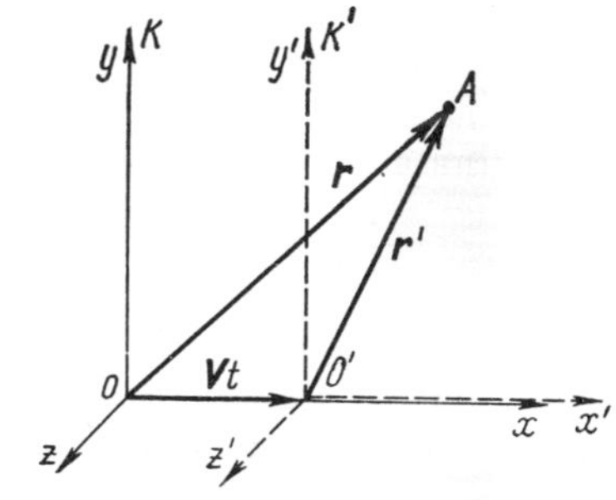

Fig. 19

This statement formulates the content of the *Galilean principle of relativity*, one of the most important principles of classical mechanics. This principle is a generalization of practice and is confirmed by all multiform applications of classical mechanics to motion of bodies whose velocity is considerably less than that of light.

Everything that was said above clearly demonstrates the exceptional nature of inertial reference frames, which as a rule makes them indispensable in studies of mechanical phenomena.

The Galilean transformation. Let us find the coordinate transformation formulae describing a transition from one inertial frame to another. Suppose the inertial frame K' moves relative to the inertial frame K with the velocity $\mathbf{V}$. Let us take the x', y', z' coordinate axes of the K' frame parallel to the respective x, y, z axes of the K frame, so that the axes x and x' coincide and are directed along the vector $\mathbf{V}$ (Fig. 19). The moment when the origins O' and O coincide is to be taken for the initial reading of time. Let us write the relation between the radius vectors $\mathbf{r}'$ and $\mathbf{r}$

of the same point A in the K' and K frames:

$$\mathbf{r}' = \mathbf{r} - \mathbf{V}t \quad (2.1)$$

and, besides,

$$t' = t. \quad (2.2)$$

The length of rods and time rate are assumed to be independent of motion here and, consequently, are identical in the two reference frames. The assumption that space and time are absolute underlies the concepts of classical mechanics, which are based on extensive experimental data pertaining to the study of motion whose velocity is substantially less than that of light.

The relations (2.1) and (2.2) are referred to as the *Galilean transformations*. These transformations can be written in a coordinate form as follows:

$$\boxed{x' = x - Vt, \quad y' = y, \quad z' = z, \quad t' = t.} \quad (2.3)$$

Differentiating Eq. (2.1) with respect to time, we get the classical law of velocity transformation for a point on transition from one inertial reference frame to another:

$$\boxed{\mathbf{v}' = \mathbf{v} - \mathbf{V}.} \quad (2.4)$$

Differentiating this expression with respect to time and taking into acocunt that $\mathbf{V} = \text{const}$, we obtain $\mathbf{w}' = \mathbf{w}$, i.e. the point accelerates equally in all inertial reference frames.

§ 2.2. The Fundamental Laws of Newtonian Dynamics

Investigating various kinds of motion in practice, we discover that in inertial reference frames *any* acceleration of a body is caused by some other bodies acting on it. The degree of influence (action) of each of the surrounding bodies on the state of motion of the body A in question is a problem whose solution in a concrete case can be obtained through experiment.

The influence of another body (or bodies) causing the acceleration of the body A is referred to as a *force*. Therefore, a body accelerates due to a force acting on it.

One of the most significant features of a force is its material origin. When speaking of a force, we always implicitly assume that in the absence of extraneous bodies the force acting on the body in question is equal to zero. If it becomes evident that a force is present, we try to identify its *origin* as one or another concrete body or bodies.

All the forces which are treated in mechanics are usually subdivided into the forces emerging due to the direct contact between bodies (forces of pressure, friction) and the forces arising due to the *fields* generated by interacting bodies (gravitational and electromagnetic forces). We should point out, however, that such a classification of forces is conditional: the interacting forces in a direct contact are essentially produced by some kind of field generated by molecules and atoms of bodies. Consequently, in the final analysis all forces of interaction between bodies are caused by fields. The analysis of the nature of interaction forces lies outside the scope of mechanics and is considered in other divisions of physics.

Mass. Experience shows that every body "resists" any effort to change its velocity, both in magnitude and direction. This property expressing the degree of unsusceptibility of a body to any change in its velocity is called *inertness*. Different bodies reveal this property in different degrees. A measure of inertness is provided by the quantity called *mass*. A body possessing a greater mass is more inert, and vice versa.

Let us introduce the notion of mass m by *defining* the ratio of masses of two different bodies via the inverse ratio of accelerations imparted to them by *equal* forces:

$$m_1/m_2 = w_2/w_1. \tag{2.5}$$

Note that this definition does not require any preliminary measurements of the forces. It is sufficient to meet the criterion of *equality* of forces. For example, if two different bodies lying on a smooth horizontal surface are pulled in succession by the same spring oriented horizontally and stretched to the same length, the influence of the spring on the bodies is equal in both cases, i.e. the force is identical in both cases.

Consequently, a comparison of the masses of two bodies experiencing the action of the same force reduces to the comparison of accelerations of these bodies. Having adopted a certain body for a mass standard, we may compare the mass of any body against the standard.

Experience shows that in terms of Newtonian mechanics a mass determined that way possesses the following two important properties:

(1) mass is an *additive* quantity, i.e. the mass of a composite body is equal to the sum of the masses of its constituents;

(2) the mass of a body proper is a *constant* quantity, remaining invariable in the process of motion.

Force. Let us get back to the experiment in which we compared the accelerations of two different bodies subjected to the action of an *equally* stretched spring. The fact that the spring was stretched equally in both cases permitted us to claim an identical force exerted by the spring.

On the other hand, a force makes a body accelerate. The accelerations of different bodies under the action of the same equally stretched spring are different. Our task is to define a force *in such a way* as to make it the same despite the difference in accelerations of different bodies in the case considered.

To do this, we have to clear up the following thing first: what quantity is the *same* in this experiment? The answer is obvious: it is the product mw. It is then natural to adopt this quantity for a definition of force. Besides, taking into account that acceleration is a vectorial quantity, we shall also assume a force to be a vector coinciding in its direction with the acceleration vector $\mathbf{w}$.

Thus, in Newtonian mechanics a *force* acting on a body of mass *m is defined* as a product $m\mathbf{w}$. Apart from the maximum simplicity and convenience, this definition of a force is of course justified only by the subsequent analysis of all consequences following from it.

Newton's second law. Examining in practice the interaction of various mass points with surrounding bodies, we observe that $m\mathbf{w}$ depends on the quantities characterizing both the state of the mass point itself and the state of surrounding bodies.

This significant physical fact underlies one of the most fundamental generalizations of Newtonian mechanics, *Newton's second law*:

the product of the mass of a mass point by its acceleration is a function of the position of this point relative to surrounding bodies, and sometimes a function of its velocity as well. This function is denoted by **F** and is called a *force.*

This is exactly what constitutes the actual content of Newton's second law, which is usually formulated in a brief form as follows:

the product of the mass of a mass point by its acceleration is equal to the force acting on it, i.e.

$$\boxed{m\mathbf{w} = \mathbf{F}.} \tag{2.6}$$

This equation is referred to as the *motion equation of a mass point.*

It should be immediately emphasized that Newton's second law and Eq. (2.6) acquire specific meaning only after the function **F** is established, that is, its dependence on the quantities involved, or the *law of force*, is known. Determining the law of force in each specific case is one of the basic problems of physical mechanics.

The definition of a force as $m\mathbf{w}$ (Eq. (2.6)) has the remarkable merit of presenting the laws of force in a very simple form. The study of motions at relativistic velocities, however, showed that the laws of force should be modified to make the forces dependent on the velocity of a mass point in an intricate way. The theory would thus turn out to be cumbersome and confusing.

However, there is an easy way to dispose of the problem; the definition of a force should be slightly modified as follows: *a force is a derivative of the momentum* **p** *of a mass point with respect to time*, that is, $d\mathbf{p}/dt$; Eq. (2.6) should then be rewritten as $d\mathbf{p}/dt = \mathbf{F}$.

In Newtonian mechanics this definition of a force is identical to $m\mathbf{w}$ since $\mathbf{p} = m\mathbf{v}$, $m = \text{const}$ and $d\mathbf{p}/dt = m\mathbf{w}$, while in relativistic mechanics, as we shall see, momentum depends on the velocity of a mass point in a more complicated way. But something different is important here. When force is defined as $d\mathbf{p}/dt$, the laws of forces prove to remain the same in the relativistic case as well. Thus, the simple expression of a given force via the physical surrounding should not be changed on transition to relativistic mechanics. This fact will be employed later.

On summation of forces. Under the given specific conditions any mass point experiences, strictly speaking, only *one* force $\mathbf{F}$ whose magnitude and direction are specified by the position of that point relative to all surrounding bodies, and sometimes by its velocity as well. And still very often it is convenient to depict this force $\mathbf{F}$ as a cumulative action of individual bodies, or a sum of the forces $\mathbf{F}_1$, $\mathbf{F}_2$, ... Experience shows that if the bodies acting as sources of force exert no influence on each other and so do not change their state in the presence of other bodies, then

$$\mathbf{F} = \mathbf{F}_1 + \mathbf{F}_2 + \ldots,$$

where $\mathbf{F}_i$ is the force which the ith body exerts on the given mass point *in the absence* of other bodies.

If that is the case, the forces $\mathbf{F}_1$, $\mathbf{F}_2$,. . . are said to obey the *principle of superposition*. This statement should be regarded as a generalization of experimental data.

Newton's third law. In all experiments involving only two bodies A and B, body A imparting acceleration to B, it turns out that B imparts acceleration to A. Hence, we come to the conclusion that the action of bodies on one another is of an *interactive* nature.

Newton postulated the following general property of all interaction forces, *Newton's third law*:

two mass points act on each other with forces which are always equal in magnitude and oppositely directed along a straight line connecting these points, i.e.

$$\boxed{\mathbf{F}_{12} = -\mathbf{F}_{21}.} \tag{2.7}$$

This implies that interaction forces always appear in *pairs*. The two forces are applied to *different* mass points; besides, they are the forces of the *same* nature.

The law (2.7) holds true for systems comprising any number of mass points. We proceed from the assumption that in this case as well the interaction reduces to the forces of paired interaction between mass points.

In Newton's third law both forces are assumed to be equal in magnitude at *any* moment of time *regardless of the motion of the points*. This statement corresponds to the Newtonian idea about the instantaneous propagation of interactions, an assumption which is identified in classical mechanics as the *principle of long-range action*. In accordance with this principle the interaction between bodies propagates in space at an infinite velocity. In other words, having changed the position (state) of one body, we can immediately detect at least a slight variation in the other bodies interacting with it, however far they may be located.

Now we know that this is actually not the case: there does exist a *finite* maximum velocity of interaction propagation, being equal to the velocity of light *in vacuo*. Accordingly, Newton's third law (as well as the second one) is valid only within certain bounds. However, in classical mechanics, treating bodies moving with velocities substantially lower than the velocity of light, both laws hold true with a very high accuracy. This is evidenced, for example, by orbits of planets and artificial satellites computed with an "astronomical" accuracy by the use of Newton's laws.

Newton's laws are the fundamental laws of classical mechanics. They make it possible, at least in principle, to solve any mechanical problem. Besides, all the other laws of classical mechanics can be derived from Newton's laws.

In accordance with the Galilean principle of relativity the laws of mechanics are identical throughout all inertial reference frames. This means, specifically, that Eq. (2.6) will have the same form in any inertial reference frame. In fact, the mass m of a mass point *per se* does not depend on velocity, i.e. is the same in all reference frames. Moreover, in all inertial reference frames the acceleration $\mathbf{w}$ of a point is also identical. The force $\mathbf{F}$ is also independent of the choice of a reference frame since it is determined only by the position and velocity of a mass point relative to surrounding bodies, and in accordance with non-relativistic kinematics these quantities are equal in different inertial reference frames.

Thus, the three quantities m, **w** and **F** appearing in Eq. (2.6) do not change on transition from one inertial reference frame to another, and therefore Eq. (2.6) does not change either. In other words, the equation $m\mathbf{w} = \mathbf{F}$ is *invariant* with respect to the Galilean transformation.

§ 2.3. Laws of Forces

In accordance with Eq. (2.6) the motion laws of a particle can be determined in strictly mathematical terms provided we know the laws of forces acting on this particle, that is, the dependence of the force on the quantities determining it. In the final analysis, each law of this kind is obtained from the processing of experimental data, and, basically, always rests on Eq. (2.6) as a definition of force.

Gravitational and electrical forces are the most fundamental forces underlying all mechanical phenomena. Let us describe briefly these forces in the simplest form when interacting masses (charges) are at rest or move with a low (non-relativistic) velocity.

The gravitational force acting between two mass points. In accordance with the *law of universal gravitation* this force is proportional to the product of the masses of points m_1 and m_2, inversely proportional to the square of the distance r between them and directed along the straight line connecting these points:

$$F = \gamma \frac{m_1 m_2}{r^2}, \tag{2.8}$$

where γ is the gravitation constant.

The masses involved in this law are called *gravitational* in distinction to *inert* masses entering Newton's second law. It was established from experience, however, that a gravitational mass and an inert mass of any body are strictly proportional to each other. Consequently, we can regard them equal (i.e. to take the same standard for measuring the two masses) and speak just of mass, whether it appears as a measure of inertness of a body or as a measure of gravitational attraction.

The Coulomb force acting between two point charges q_1 and q_2,

$$F = k\frac{q_1 q_2}{r^2}, \qquad (2.9)$$

where r is the distance between the charges and k is a proportionality constant dependent on the choice of a system of units. As distinct from the gravitational force Coulomb's force can be both attractive and repulsive.

It should be pointed out that Coulomb's law (2.9) does not hold precisely when the charges move. The electrical interaction of moving charges turns out to be dependent on their motion in a complicated way. One part of that interaction which is caused by motion is referred to as *magnetic* force (hence, another name of this interaction: the electromagnetic one). At low (non-relativistic) velocities the magnetic force constitutes a negligible part of an electric interaction, which is described by the law (2.9) with a high degree of accuracy.

In spite of the fact that gravitational and electrical interactions underlie all innumerable mechanical phenomena, the analysis of these phenomena, especially macroscopic ones, would prove to be very complicated if we proceeded in all cases from these fundamental interactions. Therefore, it is convenient to introduce some other, approximate, laws of forces which can in principle be obtained from the fundamental forces. This way we can simplify the problem in mathematical terms and to turn it into a practically soluble one.

With this in mind, the following forces can be, for example, introduced.

The uniform force of gravity

$$\mathbf{F} = m\mathbf{g}, \qquad (2.10)$$

where m is the mass of a body and $\mathbf{g}$ is gravity acceleration.*

The elastic force is proportional to a displacement of a mass point from the equilibrium position and directed to-

* Note that in contrast to the force of gravity the *weight* $\mathbf{P}$ is the force which a body exerts on a support or a suspension which is *motionless* relative to this body. For example, if a body with its support (suspension) is at rest with respect to the Earth, the weight $\mathbf{P}$ coincides with the gravity force. Otherwise, $\mathbf{P} = m(\mathbf{g} - \mathbf{w})$, where $\mathbf{w}$ is the acceleration of the body (with the support) relative to the Earth.

ward the equilibrium position:

$$\mathbf{F} = -\varkappa \mathbf{r}, \tag{2.11}$$

where **r** is the radius vector describing the displacement of a particle from the equilibrium position; and $\varkappa$ is a positive constant characterizing the "elastic" properties of a particular force. An example of such a force is that of elastic deformation arising from an extension (constriction) of a spring or a bar. In accordance with *Hooke's law* this force is defined as $F = \varkappa \Delta l$, where Δl is the magnitude of elastic deformation.

The sliding friction force, emerging when a given body slides over the surface of another body

$$F = kR_n, \tag{2.12}$$

where k is the sliding friction coefficient depending on the nature and condition of the contacting surfaces (specifically, their roughness), and R_n is the force of the normal pressure squeezing the rubbing surfaces together. The force **F** is directed oppositely to the motion of a given body relative to another body.

The resistance force acting on a body during its translation through fluid. This force depends on the velocity **v** of a body relative to a medium and is directed oppositely to the **v** vector:

$$\mathbf{F} = -k\mathbf{v}, \tag{2.13}$$

where k is a positive coefficient intrinsic to a given body and a given medium. Generally speaking, this coefficient depends on the velocity v, but in many cases at low velocities it can be regarded practically constant.

§ 2.4. The Fundamental Equation of Dynamics

The fundamental equation of dynamics of a mass point is nothing but a mathematical expression of Newton's second law:

$$\boxed{m\frac{d\mathbf{v}}{dt} = \mathbf{F}.} \tag{2.14}$$

Basically, Eq. (2.14) is a differential equation of motion of a point in vector form. Its solution constitutes the basic problem of dynamics of a mass point. Two antithetic formulations of the problem are possible here:

(1) to find the force **F** acting on a point if the mass m of the point and the time dependence of its radius vector **r** (t) are known, and

(2) to find the motion law of a point, i.e. the time dependence of its radius vector **r** (t), if the mass m of the point and the force **F** (or the forces $\mathbf{F}_i$) are known together with the initial conditions, the velocity $\mathbf{v}_0$ and the position $\mathbf{r}_0$ of the point at the initial moment of time.

In the first case the problem reduces to differentiating **r** (t) with respect to time and in the second to integrating Eq. (2.14). The mathematical aspects of this problem were discussed at length when we treated kinematics of a point.

Depending on the nature and formulation of a specific problem Eq. (2.14) is solved either in vector form or in coordinates or in projections on the tangent and on the normal to the trajectory at a given point. Let us see how Eq. (2.14) is written in the last two cases.

In projections on the Cartesian coordinate axes. Projecting both sides of Eq. (2.14) on the x, y, z axes, we get three differential equations

$$\boxed{m\frac{dv_x}{dt}=F_x, \qquad m\frac{dv_y}{dt}=F_y, \qquad m\frac{dv_z}{dt}=F_z,} \tag{2.15}$$

where F_x, F_y, F_z are the projections of the vector **F** on the x, y, z axes. It should be borne in mind that these projections are algebraic quantities: depending on the orientation of the vector **F** they may be both positive and negative. The sign of the projection of the resultant force **F** also defines the sign of the projection of the acceleration vector.

Let us show a concrete example of the *standard method* of solving problems through the use of Eq. (2.15).

Example. A small bar of mass m slides down an inclined plane forming an angle α with the horizontal. The friction coefficient is equal to k. Find the acceleration of the bar relative to the plane. (This reference frame is assumed to be inertial.)

First of all we should depict all the forces acting on the bar: the force of gravity $m\mathbf{g}$, the normal force of reaction $\mathbf{R}$ of the plane and the friction force $\mathbf{F}_{fr}$ (Fig. 20) directed oppositely to the motion of the bar.

After that let us fix the coordinate system x, y, z to the "inclined-plane" reference frame. Generally speaking, a coordinate system can be oriented at will, but in many cases (and in this one, in particular) the direction of the axes is specified by the character of motion. In this case, for example, the direction in which the bar moves is known in advance, and therefore the coordinate axes should be so laid out that one of them coincides with the motion direction. Then the problem reduces to the solution of only one of the equations (2.15). Thus, let us choose the x axis as shown in Fig. 20, and indicate its positive direction by an arrow.

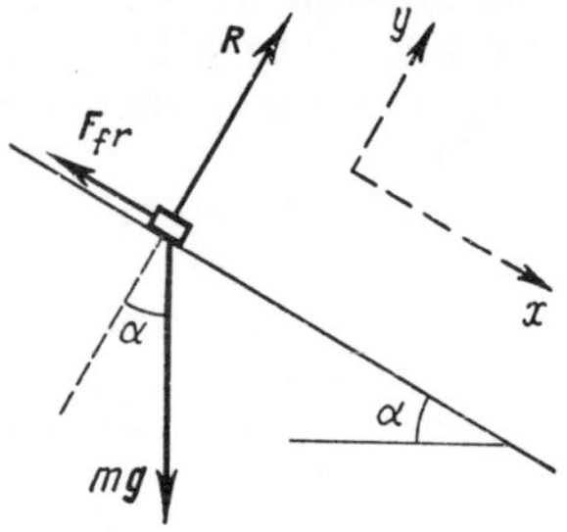

Fig. 20

And only now we can set about working out Eq. (2.15): the left-hand side contains the product of the mass m of the bar by the projection of its acceleration w_x, and the right-hand side the projections of all forces on the x axis:

$$mw_x = mg_x + R_x + F_{fr\,x}.$$

In this case $g_x = g \sin\alpha$, $R_x = 0$ and $F_{fr\,x} = -F_{fr}$, and therefore

$$mw_x = mg \sin\alpha - F_{fr}.$$

Since the bar moves only along the x axis, the sum of projections of all forces on any direction perpendicular to the x axis is equal to zero in accordance with Newton's second law. Taking the y axis as such a direction (Fig. 20), we obtain

$$R = mg \cos\alpha \quad \text{and} \quad F_{fr} = kR = kmg \cos\alpha.$$

And finally,

$$mw_x = mg \sin\alpha - kmg \cos\alpha.$$

If the right-hand side of this equation is positive, then also $w_x > 0$, and consequently the vector $\mathbf{w}$ is directed down along the inclined plane, and vice versa.

In projections on the tangent and the normal to the trajectory at a given point. Projecting both sides of Eq. (2.14) on the travelling unit vectors $\boldsymbol{\tau}$ and $\mathbf{n}$ (Fig. 21) and making use of the tangential and normal accelerations appearing in Eq. (1.10), we can write

$$\boxed{m\frac{dv_\tau}{dt} = F_\tau, \qquad m\frac{v^2}{\rho} = F_n,} \tag{2.16}$$

where F_τ and F_n are the projections of the vector **F** on the unit vectors $\boldsymbol{\tau}$ and **n**. In Fig. 21 both projections are positive. The vectors $\mathbf{F}_\tau$ and $\mathbf{F}_n$ are referred to as the tangential and normal components of the force **F**.

Recall that the unit vector $\boldsymbol{\tau}$ is oriented in the direction of growing arc coordinate l while the unit vector **n** is directed to the centre of curvature of the trajectory at a given point.

Eqs. (2.16) are convenient to use provided the trajectory of a mass point is known.

Example. A small body A slides off the top of a smooth sphere of radius r. Find the velocity of the body at the moment it loses contact with the surface of the sphere if its initial velocity is negligible.

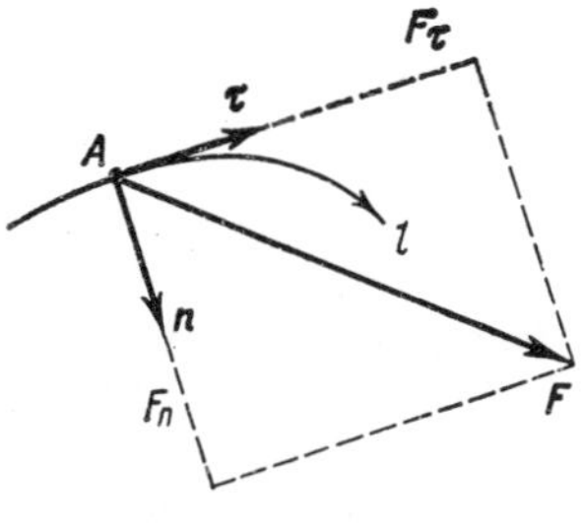

Fig. 21

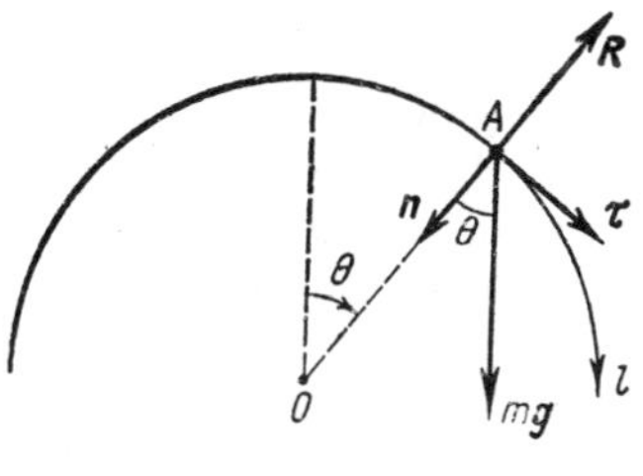

Fig. 22

Let us depict the forces acting on the body A (which are the force of gravity mg and the normal force of reaction **R**) and write Eqs. (2.16) via projections on the unit vectors $\boldsymbol{\tau}$ and **n** (Fig. 22):

$$m\,dv/dt = mg \sin \theta,$$

$$mv^2/r = mg \cos \theta - R;$$

since the subindex τ is inessential here, it has been omitted.

The first equation should be transformed to make it more convenient to integrate. Taking into consideration that $dt = dl/v = r\,d\theta/v$ where dl is an elementary path the body A covers during the time interval dt, we shall write the first equation in the following form:

$$v\,dv = gr \sin \theta \, d\theta.$$

Integrating the left-hand side of this expression between the limits 0 and v and the right-hand side between 0 and θ, we find

$$v^2 = 2gr\,(1 - \cos \theta).$$

Next, at the moment the body loses contact with the surface $R = 0$, and therefore the second initial equation takes the form

$$v^2 = gr \cos \theta.$$

where v and θ correspond to the moment when the body loses contact with the surface. Eliminating $\cos\theta$ from the last two equalities, we obtain $v = \sqrt{2gr/3}$.

§ 2.5. Non-inertial Reference Frames. Inertial Forces

The fundamental equation of dynamics in a non-inertial frame. As mentioned above, the fundamental equation of dynamics holds true only in inertial reference frames. Still there are many cases when a specific problem needs to be solved in a *non-inertial* reference frame (e.g. motion of a simple pendulum in a carriage moving with an acceleration, motion of a satellite relative to the Earth surface etc.). Hence, the following question arises: how to modify the fundamental equation of dynamics to make it valid in non-inertial reference frames?

With this in mind let us consider two reference frames: the *inertial* frame K and *non-inertial* frame K'. Suppose that we know the mass m of a particle, the force $\mathbf{F}$ exerted on this particle by surrounding bodies and the character of motion of the K' frame relative to the K frame.

Let us examine a sufficiently general case when the K' frame rotates with a constant angular velocity $\boldsymbol{\omega}$ about an axis which translates relative to the K frame with the acceleration $\mathbf{w}_0$. We shall employ the acceleration transformation formula (1.31), from which it follows that the acceleration of the particle in the K' frame is

$$\mathbf{w}' = \mathbf{w} - \mathbf{w}_0 + \omega^2\boldsymbol{\rho} + 2\,[\mathbf{v}'\boldsymbol{\omega}], \tag{2.17}$$

where $\mathbf{v}'$ is the velocity of the particle relative to the K' frame and $\boldsymbol{\rho}$ is the radius vector perpendicular to the rotation axis and describing the position of this particle with respect to this axis.

Multiplying both sides of Eq. (2.17) by the mass m of the particle and taking into account that in an inertial reference frame $m\mathbf{w} = \mathbf{F}$, we obtain

$$\boxed{m\mathbf{w}' = \mathbf{F} - m\mathbf{w}_0 + m\omega^2\boldsymbol{\rho} + 2m\,[\mathbf{v}'\boldsymbol{\omega}].} \tag{2.18}$$

This is the *fundamental equation of dynamics in a non-inertial reference frame* rotating with a constant angular velocity $\boldsymbol{\omega}$

about an axis translating with the acceleration $\mathbf{w}_0$. It indicates that even if $\mathbf{F} = 0$, the particle will move in this frame with an acceleration (which in the general case differs from zero), as if under the influence of certain forces corresponding to the last three terms of Eq. (2.18). These forces are referred to as *inertial.*

Eq. (2.18) shows that the introduction of inertial forces makes it possible to keep the format of the fundamental equation of dynamics in non-inertial reference frames as well: the left-hand side is the product of the mass of the particle by its acceleration (but this time relative to the non-inertial reference frame), and the right-hand side contains the forces. However, apart from the force **F** caused by the influence of surrounding bodies (interaction forces), it is necessary to take into account inertial forces (the remaining terms on the right-hand side of Eq. (2.18)).

Inertial forces. Let us write Eq. (2.18) in the following form:

$$m\mathbf{w}' = \mathbf{F} + \mathbf{F}_{in} + \mathbf{F}_{cf} + \mathbf{F}_{Cor}, \tag{2.19}$$

where

$$\boxed{\mathbf{F}_{in} = -m\mathbf{w}_0} \tag{2.20}$$

is the inertial force caused by the translation of the non-inertial reference frame;

$$\boxed{\mathbf{F}_{cf} = m\omega^2\boldsymbol{\rho}} \tag{2.21}$$

is the *centrifugal force of inertia*;

$$\boxed{\mathbf{F}_{Cor} = 2m\,[\mathbf{v}'\boldsymbol{\omega}]} \tag{2.22}$$

is the *Coriolis force.* The last two forces emerge due to rotation of the reference frame.

Thus, we see that the inertial forces depend on the characteristics of the non-inertial reference frame ($\mathbf{w}_0$, $\boldsymbol{\omega}$) as well as on the distance $\boldsymbol{\rho}$ and the velocity $\mathbf{v}'$ of a particle in that reference frame.

For example, if a non-inertial reference frame translates relative to an inertial one, a free particle in that frame experiences only the force (2.20) whose direction is opposite to the acceleration $\mathbf{\omega}_0$ of the given reference frame. Recall how a sudden braking of the carriage we travel in makes us swing forward, that is, in the direction opposite to $\mathbf{w}_0$.

Here is another example: a reference frame rotates about a stationary axis with the angular velocity $\mathbf{\omega}$, and the body A is at rest in that frame (e.g. you are on a rotating platform in an amusement park). Apart from the forces of interaction with surrounding bodies, the body A experiences the centrifugal force of inertia (2.21) directed along the radius vector $\boldsymbol{\rho}$ from the rotation axis. As long as the body A is at rest relative to the rotating platform ($\mathbf{v}' = 0$), this force makes up for the interaction force. But as soon as the body begins to move, i.e. the velocity $\mathbf{v}'$ appears, there originates the Coriolis force (2.22) whose direction is determined by the vector cross product $[\mathbf{v}'\boldsymbol{\omega}]$. Note that the Coriolis force crops up to supplement the centrifugal force of inertia appearing irrespective of whether the body is at rest or moves with respect to the rotating reference frame.

It was pointed out that the reference frame fixed to the Earth's surface can be regarded in many cases as practically inertial. However, there are some phenomena whose interpretation in this reference frame is impossible unless its non-inertial nature is taken into account.

For instance, free-fall acceleration is the greatest at the Earth's poles. Approaching the equator, one observes a decrease in this acceleration caused not only by the deviations of the Earth from a spherical shape, but also by the growing action of the centrifugal force of inertia. There are also such phenomena as a deviation of free-falling bodies to the East, a wash-out of right banks of rivers in the Northern Hemisphere and left banks in the Southern Hemisphere, a rotation of the Foucault pendulum oscillation plane, etc. Phenomena of this kind are associated with the motion of bodies relative to the Earth's surface and can be explained by the Coriolis force.

Example. A train of mass m moves along a meridian at the latitude φ with the velocity $\mathbf{v}'$. Find the lateral force which the train exerts on the rails.

In the reference frame fixed to the Earth (rotating at the angular velocity $\boldsymbol{\omega}$) the train's acceleration component normal to the meridian plane is equal to zero. Therefore, the sum of the projections of forces acting on the train in this direction is also equal to zero. And this means that the Coriolis force $\mathbf{F}_{Cor}$ (Fig. 23) must be counterbalanced by the lateral force **R** exerted by the right rail on the train, i.e. $\mathbf{F}_{Cor} = -\mathbf{R}$. In accordance with Newton's third law the train acts on that rail in the horizontal direction with the force $\mathbf{R}' = -\mathbf{R}$.

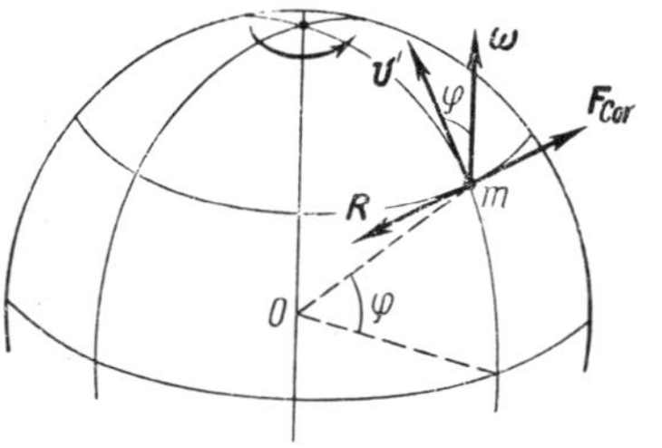

Fig. 23

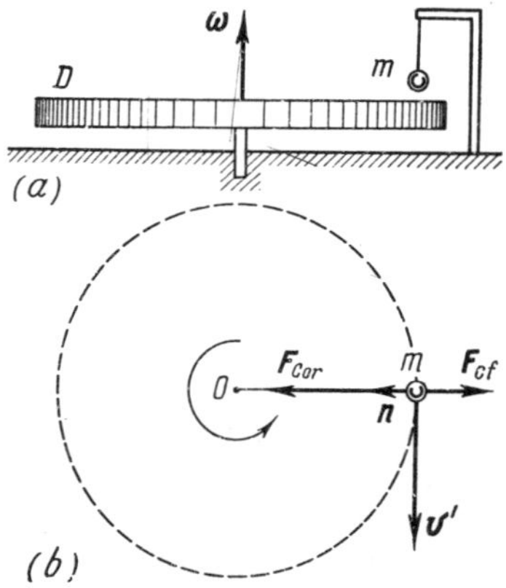

Fig. 24

Consequently, $\mathbf{R}' = \mathbf{F}_{Cor} = 2\,m\,[\mathbf{v}'\boldsymbol{\omega}]$. The magnitude of the vector $\mathbf{R}'$ is equal to $R' = 2mv'\,\omega \sin \varphi$.

The following simple example illustrates how the inertial forces "appear" on transition from an inertial reference frame to a non-inertial one.

Example. A horizontal disc D freely rotates over the surface of a table about a vertical axis with a constant angular velocity ω. A sphere possessing mass m is suspended over the disc as shown in Fig. 24*a*. Let us consider the behaviour of that sphere in the K frame fixed to the table and assumed inertial, and in the K' frame fixed to the rotating disc.

In the inertial K frame the sphere is subjected to two forces, the gravity force and the stretching force of the thread. These forces equalize each other so that the sphere is at rest in the K frame.

In the non-inertial K' frame the sphere moves uniformly along a circle with the normal acceleration $\omega^2\rho$, where ρ is the distance between the sphere and the rotation axis. One can easily see that this acceleration is due to inertial forces. Indeed, in the K' frame, apart from the two counterbalancing forces mentioned above, there are also the centripetal force of inertia and the Coriolis force (Fig. 24*b*). Taking the projections of these forces on the normal **n** to the path at the point where the sphere is located, we write:

$$mw'_n = F_{Cor} - F_{cf} = 2mv'\omega - m\omega^2\rho = m\omega^2\rho,$$

where it is taken into consideration that in this case $v' = \omega\rho$. Hence, $w'_n = \omega^2\rho$.

Properties of inertial forces. To summarize, we shall list the most significant properties of these forces in order to discriminate them from interaction forces:

1. Inertial forces are caused not by the interaction of bodies, but by the properties of non-inertial reference frames themselves. Therefore inertial forces do not obey Newton's third law.

2. To avoid misunderstandings is should be firmly borne in mind that these forces exist only in non-inertial reference frames. In inertial reference frames there are no inertial forces at all, and the notion of *force* is employed in these frames only in the Newtonian sense, that is, as a measure of interaction of bodies.

3. Just as gravitational forces, all inertial forces are proportional to the mass of a body. Consequently, in a uniform field of inertial forces, as in the field of gravitational forces, all bodies move with the same acceleration regardless of their masses. This highly important fact has far-reaching consequences.

The principle of equivalence. Since inertial forces, just as gravitational ones, are proportional to the masses of bodies, the following important conclusion can be made. Suppose we are in a certain closed laboratory and are deprived of observing the external world. Moreover, let us assume that we are not aware of the whereabouts of our laboratory: outer space or, e.g. the Earth. Observing the bodies falling with an equal acceleration regardless of their masses, we cannot determine the cause of this acceleration from only this fact. The acceleration can be brought about by a gravitational field, by an accelerated translation of the laboratory itself, or by both causes. In such a laboratory no experiment whatsoever on free fall of bodies can distinguish the uniform field of gravitation from the uniform field of inertial forces.

Einstein argued that no physical experiments of any kind can be of use to distinguish the uniform field of gravitation from the uniform field of inertial forces. This suggestion;

raised to a postulate, provides a basis for the so-called *principle of equivalence* of gravitational and inertial forces: *all physical phenomena proceed in the uniform field of gravitation in exactly the same way as in the corresponding uniform field of inertial forces.*

This far-reaching analogy between gravitational and inertial forces was used by Einstein as a starting point in his development of the *general theory of relativity*, or the relativistic theory of gravitation.

In conclusion it should be pointed out that any mechanical problem can be solved in both inertial and non-inertial reference frames. Usually the choice of one or another reference frame is determined by the formulation of the problem or by the desire to solve it in as straightforward a manner as possible. In so doing, we quite often find that non-inertial reference frames are most convenient to apply (see Problems 2.9-2.11).

Problems to Chapter 2

●**2.1.** A bar of mass m_1 is placed on a plank of mass m_2, which rests on a smooth horizontal plane (Fig. 25). The coefficient of friction between the surfaces of the bar and the plank is equal to k. The

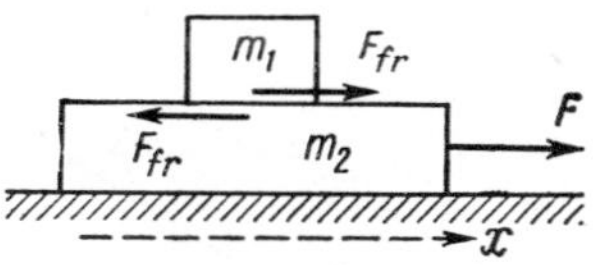

Fig. 25

plank is subjected to the horizontal force F depending on time t as $F = at$ (a is a constant). Find:

(1) the moment of time t_0 at which the plank starts sliding from under the bar;

(2) the accelerations of the bar w_1 and of the plank w_2 in the process of their motion.

Solution. 1. Let us write the fundamental equation of dynamics for the plank and the bar, having taken the positive direction of the x axis as shown in the figure:

$$m_1 w_1 = F_{fr}, \quad m_2 w_2 = F - F_{fr}. \tag{1}$$

As the force F grows, so does the friction force F_{fr} (at the initial moment it represents the friction of rest). However, the friction force

F_{fr} has the ultimate value $F_{fr.\,max} = km_1g$. Unless this value is reached, both bodies move as a single whole with equal accelerations. But as soon as the force F_{fr} reaches the limit, the plank starts sliding from under the bar, i.e.

$$w_2 \geqslant w_1.$$

Substituting here the values of w_1 and w_2 taken from Eq. (1) and taking into account that $F_{fr} = km_1g$, we obtain

$$(at - km_1g)/m_2 \geqslant kg,$$

where the sign "=" corresponds to the moment $t = t_0$. Hence,

$$t_0 = (m_1 + m_2)\, kg/a.$$

2. If $t \leqslant t_0$, then

$$w_1 = w_2 = at/(m_1 + m_2);$$

and if $t \geqslant t_0$, then

$$w_1 = kg = \text{const}, \qquad w_2 = (at - km_1g)/m_2.$$

The plots $w_1(t)$ and $w_2(t)$ are shown in Fig. 26.

●2.2. In the arrangement of Fig. 27 the inclined plane forms the angle $\alpha = 30°$ with the horizontal. The ratio of the masses shown is $\eta = m_1/m_2 = 2/3$. The coefficient of friction between the plane and

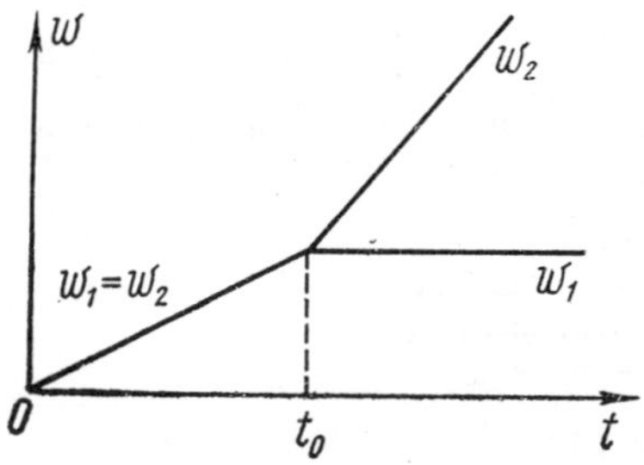

Fig. 26

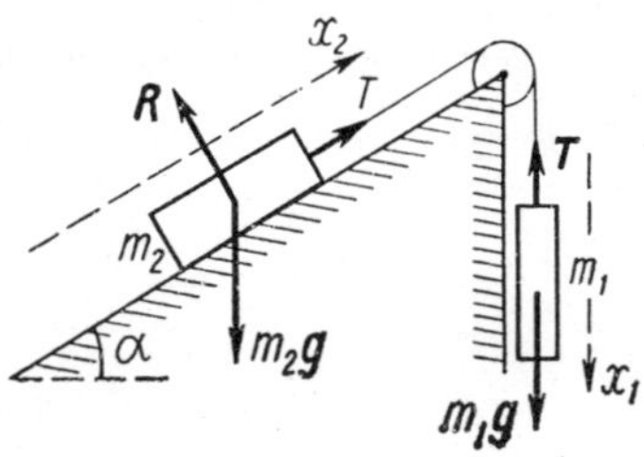

Fig. 27

the body m_2 is $k = 0.10$. The masses of the pulley block and the threads are negligible. Find the magnitude and direction of the acceleration of the body m_1 if the system is set into motion from an initial state of rest.

Solution. First we should tackle the problem associated with the direction of the friction force acting on the body m_2. Otherwise, we cannot write the fundamental equation of dynamics for the body m_2 in terms of projections, and the problem proves to be uncertain.

We shall argue as follows: suppose that in the absence of friction forces the body m_2 starts sliding, say, upward along the inclined plane. "Switching on" the friction forces, we obviously cannot reverse the motion direction, but only decrease the acceleration. Thus, the direction of the friction force acting on the body m_2 is determined if we find the acceleration direction of this body in the absence of friction ($k = 0$). Accordingly, we shall begin with that.

Let us write the fundamental equation of dynamics for both bodies in terms of projections, having taken the positive directions of the x_1 and x_2 axes as shown in Fig. 27:

$$m_1 w_x = m_1 g - T, \quad m_2 w_x = T - m_2 g \sin \alpha,$$

where T is the tensile force of the thread. Summing up termwise the left- and right-hand sides of these equations, we obtain

$$w_x = \frac{\eta - \sin \alpha}{\eta + 1} g.$$

After the substitution $\eta = 2/3$ and $\alpha = 30°$ this expression yields $w_x > 0$. i.e. the body m_2 moves up the inclined plane. Consequently, the friction force acting on this body is directed oppositely. Taking this into account, we again write the equations of motion:

$$m_1 w'_x = m_1 g - T', \quad m_2 w'_x = T' - m_2 g \sin \alpha - k m_2 g \cos \alpha.$$

Hence,

$$w'_x = \frac{\eta - \sin \alpha - k \cos \alpha}{\eta + 1} g \approx 0.05\ g.$$

●**2.3.** A non-stretchable thread with masses m_1 and m_2 attached to its ends ($m_1 > m_2$) is thrown over a pulley block (Fig. 28). We begin to lift the pulley block with the acceleration $\mathbf{w}_0$ relative to the Earth. Assuming the thread to slide over the pulley block without friction, find the acceleration w_1 of the mass m_1 relative to the Earth.

Solution. Let us designate the positive direction of the x axis as shown in Fig. 28 and write the fundamental equation of dynamics for the two masses in terms of projections on this axis:

$$m_1 w_{1x} = T - m_1 g, \tag{1}$$

$$m_2 w_{2x} = T - m_2 g. \tag{2}$$

These two equations contain three unknown quantities: w_{1x}, w_{2x}, and T. The third equation is provided by the kinematic relationship between the accelerations:

$$\mathbf{w}_1 = \mathbf{w}_0 + \mathbf{w}', \quad \mathbf{w}_2 = \mathbf{w}_0 - \mathbf{w}',$$

where $\mathbf{w}'$ is the acceleration of the mass m_1 with respect to the pulley block. Summing up termwise the left-hand and the right-hand sides of these equations, we get

$$\mathbf{w}_1 + \mathbf{w}_2 = 2\mathbf{w}_0,$$

or in terms of projections on the x axis

$$w_{1x} + w_{2x} = 2w_0. \tag{3}$$

The simultaneous solution of Eqs. (1), (2) and (3) yields

$$w_{1x} = [2m_2 w_0 + (m_2 - m_1) g]/(m_1 + m_2).$$

Whence it is seen that for a given w_0 the sign of w_{1x} depends on the ratio of the masses m_1 and m_2.

●**2.4.** A small disc moves along an inclined plane whose friction coefficient $k = \tan\alpha$, where α is the angle which the plane forms

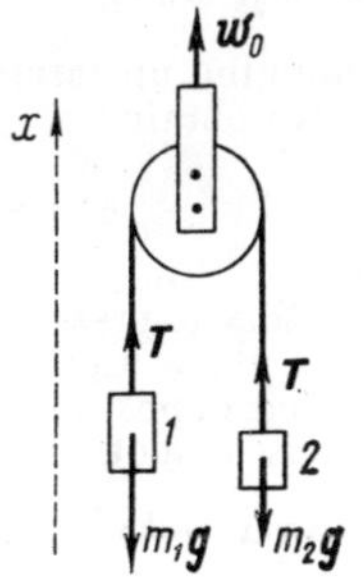

Fig. 28

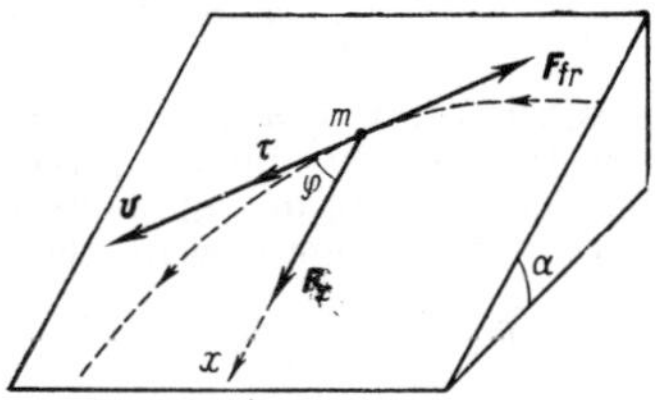

Fig. 29

with the horizontal. Find how the velocity v of the disc depends on the angle φ between the vector **v** and the x axis (Fig. 29) if at the initial moment $v = v_0$ and $\varphi = \pi/2$.

Solution. The acceleration of the disc along the plane is determined by the projection of the force of gravity on this plane $F_x = mg \sin\alpha$ and the friction force $F_{fr} = kmg \cos\alpha$. In our case $k = \tan\alpha$ and therefore

$$F_{fr} = F_x = mg \sin\alpha.$$

Let us find the projections of the acceleration on the direction of the tangent to the trajectory and on the x axis:

$$mw_\tau = F_x \cos\varphi - F_{fr} = mg \sin\alpha (\cos\varphi - 1),$$
$$mw_x = F_x - F_{fr} \cos\varphi = mg \sin\alpha (1 - \cos\varphi).$$

It is seen from this that $w_\tau = -w_x$, which means that the velocity v and its projection v_x differ only by a constant value C which does not change with time, i.e.

$$v = -v_x + C,$$

where $v_x = v \cos\varphi$. The constant C is found from the initial condition $v = v_0$, whence $C = v_0$. Finally we obtain

$$v = v_0/(1 + \cos\varphi).$$

In the course of time $\varphi \to 0$ and $v \to v_0/2$.

● **2.5.** A thin uniform elastic cord of mass m and length l_0 (in a non-stretched state) has a coefficient of elasticity $\varkappa$. After having the ends of the cord spliced, it was placed on a smooth horizontal plane, shaped as a circle and set into rotation with the angular velocity ω about the vertical axis passing through the centre of the circle. Find the tension of the cord in this state.

Solution. Let us single out a small element of the cord of mass δm as shown in Fig. 30*a*. This element moves along the circle due

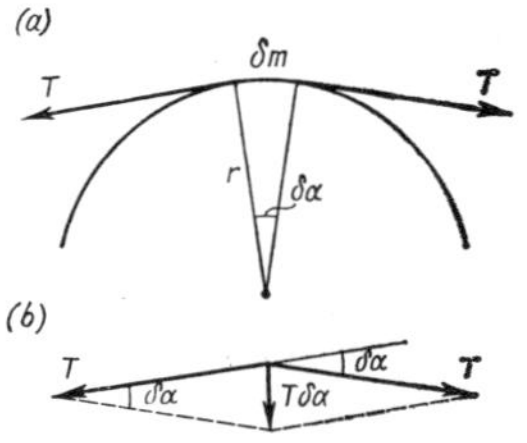

Fig. 30

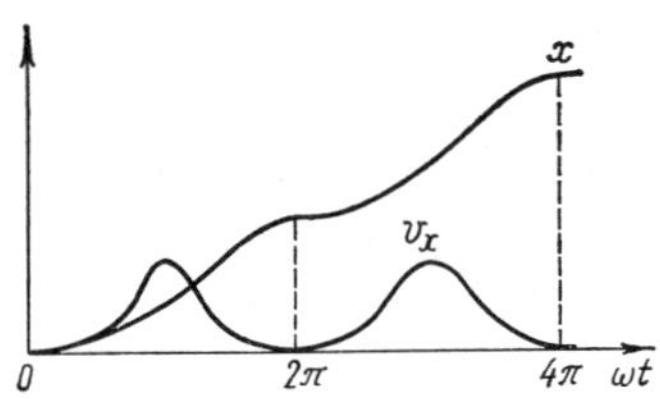

Fig. 31

to a force which is a geometric sum of two vectors each of which has the magnitude of the tension sought T (Fig. 30*b*). Consequently, in accordance with Newton's second law

$$\delta m \cdot \omega^2 r = T \cdot \delta\alpha. \tag{1}$$

Since $\delta m = (m/2\pi)\,\delta\alpha$ and $r = l/2\pi$ (where l is the length of the cord in the state of rotation), Eq. (1) takes the form

$$m\omega^2 \cdot l/4\pi^2 = T. \tag{2}$$

On the other hand, in accordance with Hooke's law

$$T = \varkappa\,(l - l_0). \tag{3}$$

Eliminating l from Eqs. (2) and (3), we obtain

$$T = \frac{\varkappa l_0}{4\pi^2\varkappa/m\omega^2 - 1}.$$

Note that in the case of a non-stretchable cord ($\varkappa = \infty$) $T = m\omega^2 l_0/4\pi^2$.

● **2.6. Integration of motion equations.** A particle of mass m moves due to the action of the force $\mathbf{F}$. The initial conditions, that is, its radius vector $\mathbf{r}\,(0)$ and velocity $\mathbf{v}\,(0)$ at the moment $t = 0$, are known. Find the position of the particle as a function of time if

(1) $\mathbf{F} = \mathbf{F}_0 \sin \omega t$, $\quad \mathbf{r}\,(0) = 0$, $\quad \mathbf{v}\,(0) = 0$;

(2) $\mathbf{F} = -k\mathbf{v}$, $\quad \mathbf{r}\,(0) = 0$, $\quad \mathbf{v}\,(0) = \mathbf{v}_0$;

(3) $\mathbf{F} = -\varkappa\mathbf{r}$, $\quad \mathbf{r}\,(0) = \mathbf{r}_0$, $\quad \mathbf{v}\,(0) = \mathbf{v}_0$, with $\mathbf{v}_0 \| \mathbf{r}_0$.

Here $\mathbf{F}_0$ is a constant vector, and ω, k, $\varkappa$ are positive constants.

5—0539

Solution. 1. In accordance with the fundamental equation of dynamics the acceleration is

$$d\mathbf{v}/dt = (\mathbf{F}_0/m)\sin \omega t.$$

We obtain the elementary increment of the velocity vector $d\mathbf{v}$ during the time dt and then the increment of this vector during the time from 0 to t:

$$\mathbf{v}(t) - \mathbf{v}(0) = (\mathbf{F}_0/m)\int_0^t \sin \omega t \, dt.$$

Taking into account that $\mathbf{v}(0) = 0$, we obtain after integration

$$\mathbf{v}(t) = (\mathbf{F}_0/m\omega)(1 - \cos \omega t).$$

Now let us find the elementary displacement $d\mathbf{r}$, or the increment of the radius vector $\mathbf{r}$ of the particle during the interval dt: $d\mathbf{r} = \mathbf{v}(t)\,dt$. The increment of the radius vector during the time from 0 to t is equal to

$$\mathbf{r}(t) - \mathbf{r}(0) = (\mathbf{F}_0/m\omega)\int_0^t (1 - \cos \omega t)\, dt.$$

Integrating this expression and taking into account that $\mathbf{r}(0) = 0$, we get

$$\mathbf{r}(t) = (\mathbf{F}_0/m\omega^2)(\omega t - \sin \omega t).$$

Fig. 31 illustrates the plots $v_x(t)$ and $x(t)$, the time dependences of projections of the vectors $\mathbf{v}$ and $\mathbf{r}$ on the x axis chosen in the particle motion direction, i.e. in the direction of the $\mathbf{F}_0$ vector.

2. In this case the acceleration is

$$d\mathbf{v}/dt = -(k/m)\,\mathbf{v}.$$

To integrate this equation we must pass to the scalar form, that is, to the modulus of the vector $\mathbf{v}$:

$$dv/v = -(k/m)\,dt.$$

Integration of this equation with allowance made for the initial conditions yields: $\ln(v/v_0) = -(k/m)\,t$. After taking antilogarithms we return to the vector form:

$$\mathbf{v} = \mathbf{v}_0 e^{-kt/m}.$$

Integrating the last equation once more (and again taking into account the initial conditions), we obtain

$$\mathbf{r} = \int_0^t \mathbf{v}\,dt = (m\mathbf{v}_0/k)(1 - e^{-kt/m}).$$

Fig. 32 shows the plots of the velocity v and the path covered s as functions of time t (in our case $s = r$).

3. In this case the particle moves along the straight line coinciding with the radius vector **r**. Choosing the x axis in this direction, we can immediately write the fundamental equation of dynamics in terms of the projection on this axis:

$$\ddot{x} + \omega^2 x = 0, \tag{1}$$

where $\ddot{x}$ is the second derivative of the coordinate with respect to time, i.e. the projection of the acceleration vector, $\omega^2 = \varkappa/m$. Eq. (1) is referred to as the *equation of harmonic vibrations.*

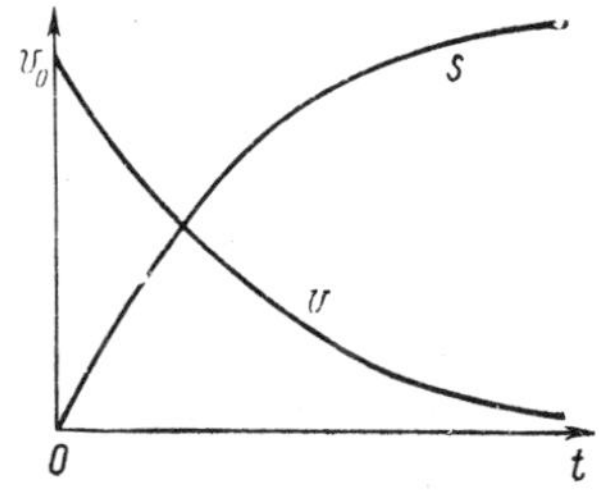

Fig. 32

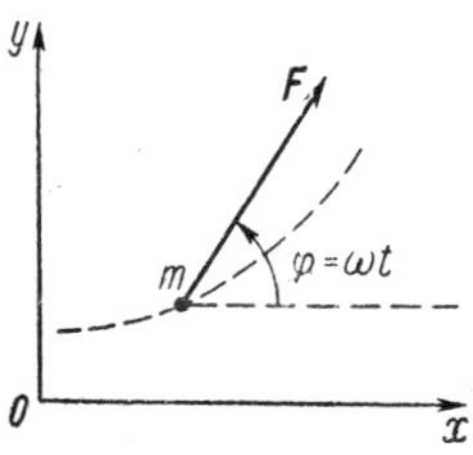

Fig. 33

It can be shown mathematically that the general solution of this equation takes the form

$$x(t) = A \cos \omega t + B \sin \omega t, \tag{2}$$

where A and B are arbitrary constants. The restrictions imposed on these constants are usually determined from the initial conditions. For instance, in our case at the moment $t = 0$

$$x(0) = x_0 \quad \text{and} \quad v_x(0) = v_{0x}, \tag{3}$$

where x_0 and v_{0x} are the projections of the $\mathbf{r}_0$ and $\mathbf{v}_0$ vectors on the x axis. After substituting Eq. (2) into Eq. (3) we get: $A = x_0$, $B = v_{0x}/\omega$. All the rest is obvious.

● 2.7. A particle of mass m moves in a certain plane due to the force **F** whose magnitude is constant and whose direction rotates with the constant angular velocity ω in that plane. At the moment $t = 0$ the velocity of the particle is equal to zero. Find the magnitude of the velocity of the particle as a function of time and the distance that the particle covers between two consecutive stops.

Solution. Let us fix the x, y coordinate system to the given plane (Fig. 33), taking the x axis in the direction along which the force vector was oriented at the moment $t = 0$. Then the fundamental equation of dynamics expressed via the projections on the x and y

axes takes the form

$$m\, dv_x/dt = F \cos \omega t, \quad m\, dv_y/dt = F \sin \omega t.$$

Integrating these equations with respect to time with allowance made for the initial condition $v(0) = 0$, we obtain

$$v_x = (F/m\omega) \sin \omega t, \quad v_y = (F/m\omega)(1 - \cos \omega t).$$

The magnitude of the velocity vector is equal to

$$v = \sqrt{v_x^2 + v_y^2} = (2F/m\omega) \sin (\omega t/2).$$

It is seen from this that the velocity v turns into zero after the time interval Δt, which can be found from the relation $\omega\, \Delta t/2 = \pi$. Consequently, the sought distance is

$$s = \int_0^{\Delta t} v\, dt = 8F/m\omega^2.$$

● **2.8.** An automobile moves with the constant tangential acceleration w_τ along the horizontal plane circumscribing a circle of radius R. The coefficient of friction between the wheels of the automobile and the surface is equal to k. What distance s will be covered by the automobile without slipping in the case of zero initial velocity?

Solution. As the velocity increases, so do both the normal and the total acceleration of the automobile. There is no slipping as long as the total acceleration required is provided by the friction force. The maximum possible value of that force $F_{max} = kmg$, where m is the mass of the automobile. Therefore, in accordance with the fundamental equation of dynamics, $m\mathbf{w} = \mathbf{F}$, the maximum value of the total acceleration is

$$w_{max} = kg. \tag{1}$$

On the other hand,

$$w_{max} = \sqrt{w_\tau^2 + (v^2/R)^2}, \tag{2}$$

where v is the velocity of the automobile at the moment its acceleration reaches the maximum value. This velocity and the sought distance s are interrelated by the following formula:

$$v^2 = 2w_\tau s. \tag{3}$$

Eliminating v and w_{max} from Eqs. (1), (2) and (3), we obtain

$$s = (R/2) \sqrt{(kg/w_\tau)^2 - 1}.$$

It is not difficult to see that the solution is meaningful only if the radicand is positive, i.e. if $w_\tau < kg$.

● **2.9. Non-inertial reference frames.** A satellite moves in the Earth's equatorial plane along a circular orbit of radius r in the west-east direction. Disregarding the acceleration due to the Earth's motion around the Sun, find the acceleration w' of the satellite in the reference frame fixed to the Earth.

Solution. Suppose K is an inertial reference frame in which the Earth's rotation axis is motionless, and K' is a non-inertial reference frame fixed to the Earth and rotating with the angular velocity ω with respect to the K frame.

To derive the acceleration w' of the satellite in the K' frame we must first of all depict all the forces acting on the satellite in that reference frame: the gravity $\mathbf{F}$, the Coriolis force $\mathbf{F}_{Cor}$ and the centrifugal force $\mathbf{F}_{cf}$ (Fig. 34, the view from the Earth's North Pole).

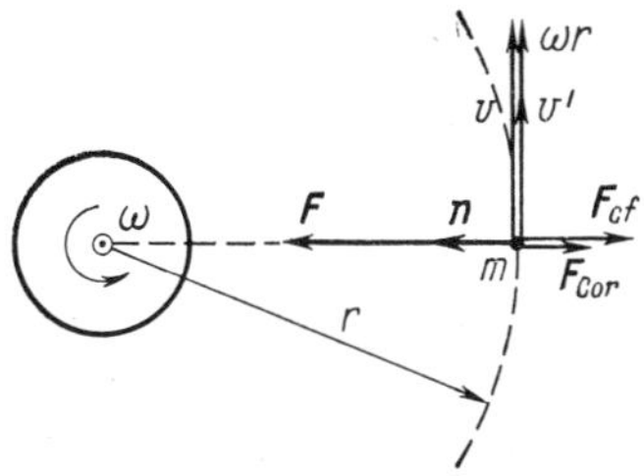

Fig. 34

Now let us make use of Eq. (2.18), assuming $\mathbf{w}_0 = 0$ (in accordance with the conditions of the problem). Since in the K' frame the satellite travels along a circle, Eq. (2.18) can be immediately written via projections on the trajectory's normal $\mathbf{n}$:

$$mw' = F - 2mv'\omega - m\omega^2 r, \quad (1)$$

where $F = \gamma mM/r^2$, and m and M are the masses of the satellite and the Earth respectively. Now we have only to find the velocity v' of the satellite in the K' frame. To do this, we shall make use of the kinematic relation (1.24) in scalar form

$$v' = v - \omega r, \quad (2)$$

where v is the velocity of the satellite in the K frame (Fig. 34), and of the equation of motion of the satellite in the K frame

$$mv^2/r = \gamma mM/r^2, \quad (3)$$

from which v is found. Solving simultaneously Eqs. (1), (2) and (3), we obtain

$$w' = \left(1 - \omega r\sqrt{r/\gamma M}\right)^2 \gamma M/r^2.$$

Specifically, $w' = 0$ when $r = \sqrt[3]{\gamma M/\omega^2} = 4.2\cdot 10^3$ km. Such a satellite is called *stationary*: it is motionless relative to the Earth's surface.

● **2.10.** A small sleeve of mass m slides freely along a smooth horizontal shaft which rotates with the constant angular velocity $\boldsymbol{\omega}$ about a fixed vertical axis passing through one of the shaft's ends. Find the horizontal component of the force which the shaft exerts on the sleeve when it is at the distance r from the axis. At the initial moment the sleeve was next to the axis and possessed a negligible velocity.

Solution. Let us examine the motion of the sleeve in a rotating reference frame fixed to the shaft. In this reference frame the sleeve moves rectilinearly. This means that the sought force is balanced out by the Coriolis force (Fig. 35):

$$\mathbf{R} = -\mathbf{F}_{Cor} = 2m\,[\boldsymbol{\omega}\mathbf{v}'] . \quad (1)$$

Thus, the problem reduces to determining the velocity $\mathbf{v}'$ of the sleeve relative to the shaft. In accordance with Eq. (2.19)

$$dv'/dt = F_{cf}/m = \omega^2 r.$$

Taking into account that $dt = dr/v'$, the last equation can be transformed to

$$v'\,dv' = \omega^2 r\,dr.$$

Integrating this equation with allowance made for the initial conditions ($v' = 0$, $r = 0$), we find $v' = \omega r$, or in a vector form

$$\mathbf{v}' = \omega \mathbf{r}. \quad (2)$$

Substituting Eq. (2) into Eq. (1), we get

$$\mathbf{R} = 2m\omega\,[\boldsymbol{\omega}\mathbf{r}].$$

● **2.11. The stability of motion.** A wire ring of radius r rotates with the constant angular velocity ω about the vertical axis OO'

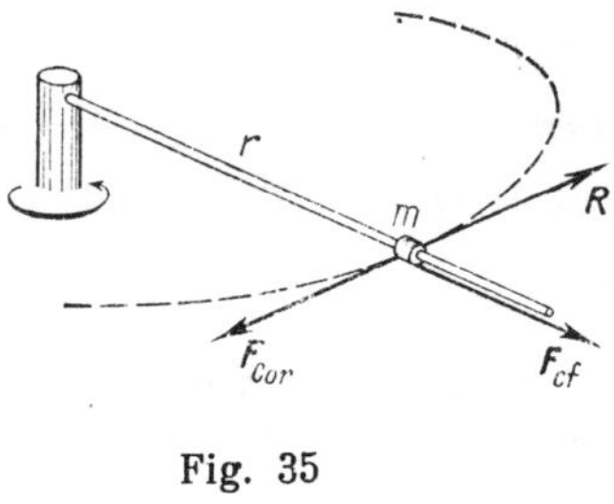

Fig. 35

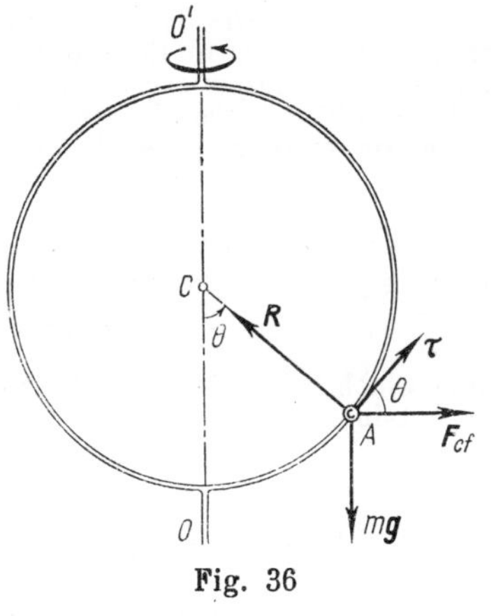

Fig. 36

passing through its diameter. A small sleeve A can slide along the ring without friction. Find the angle θ (Fig. 36) corresponding to the stable position of the sleeve.

Solution. Let us examine the behaviour of the sleeve in a reference frame fixed to the rotating ring. Its motion along the ring is characterized by the resultant force projection F_τ on the unit vector $\boldsymbol{\tau}$ at the point A. It is seen from Fig. 36 that

$$F_\tau = F_{cf} \cos\theta - mg \sin\theta.$$

The right-hand side of this equation contains the projections of the centrifugal force and gravity. Taking into account[1] that $F_{cf} = m\omega^2 r \sin\theta$, we rewrite the foregoing expression as follows:

$$F_\tau \sim \sin\theta (\cos\theta - g/\omega^2 r). \qquad (1)$$

From the equilibrium condition ($F_\tau = 0$) we can find the two values of the angle θ_0 ensuring that equilibrium: $\sin\theta_0 = 0$ and $\cos\theta_0 = g/\omega^2 r$. The first condition can be satisfied for any value of ω, while the second one only if $g/\omega^2 r < 1$. Thus, in the case of low ω values there is only one equilibrium position, at the bottom point ($\theta_0 = 0$); but in the case of large ω values ($\omega > \sqrt{g/r}$) another equilibrium position, defined by the second condition, is possible.

A certain equilibrium position is stable provided the force F_τ appearing on withdrawal of the sleeve from that position (in any direction) is directed back, to the equilibrium position, that is, the sign of F_τ must be opposite to that of the deflection $\Delta\theta$ from the equilibrium angle θ_0.

At low deflections $d\theta$ from the θ_0 angle the appearing force δF_τ may be found as a differential of expression (1):

$$\delta F_\tau \sim [\cos\theta_0 (\cos\theta_0 - g/\omega^2 r) - \sin^2\theta_0]\, d\theta.$$

At the bottom equilibrium position ($\theta_0 = 0$)

$$\delta F_\tau \sim (1 - g/\omega^2 r)\, d\theta. \qquad (2)$$

This equilibrium position is stable provided the expression put in parentheses is negative, i.e. when $\omega < \sqrt{g/r}$.

At the other equilibrium position ($\cos\theta_0 = g/\omega^2 r$)

$$\delta F_\tau \sim -\sin^2\theta_0\, d\theta.$$

It is seen that this equilibrium position (if it exists) is always stable.

Thus, as long as there is only the bottom equilibrium position (with $\omega < \sqrt{g/r}$), it is always stable. However, on the appearance of the other equilibrium position (when $\omega > \sqrt{g/r}$) the bottom position becomes unstable (see Eq. (2)), and the sleeve immediately passes from the lower to the upper position, which is always stable.

CHAPTER 3

ENERGY CONSERVATION LAW

§ 3.1. On Conservation Laws

Any body (or an assembly of bodies) represents, in fact, a system of mass points, or particles. If a system changes in the course of time, it is said that its *state* varies. The state of a system is defined by specifying the concurrent coordinates and velocities of all constituent particles.

Experience shows that if the laws of forces acting on a system's particles and the state of the system at a certain initial moment are known, the motion equations can help predict the subsequent behaviour of the system, i.e. find its state at any moment of time. That is how, for example, the problem of planetary motion in the solar system has been solved.

However, an analysis of a system's behaviour by the use of the motion equations requires so much effort (e.g. due to the complexity of the system itself), that a comprehensive solution seems to be practically impossible. Moreover, such an approach is absolutely out of the question if the laws of acting forces are not known. Besides, there are some problems in which the accurate consideration of motion of individual particles is meaningless (e.g. gas).

Under these circumstances the following question naturally comes up: are there any general principles following from Newton's laws that would help avoid these difficulties by opening up some new approaches to the solution of the problem.

It appears that such principles exist. They are called *conservation laws.*

As it was mentioned, the state of a system varies in the course of time as that system moves. However, there are some quantities, state functions, which possess the very important and remarkable property of retaining their values constant with time. Among these constant quantities, *energy*, *momentum* and *angular momentum* play the most significant role. These three quantities have the important general property of additivity: their value for a system composed of parts whose interaction is negligible is equal to the sum of the corresponding values for the individual constituent parts (incidentally, in the case of momentum and angular momentum additivity holds true even in the presence of interaction). It is additivity that makes these three quantities extremely important.

Later on it became known that the laws of conservation of energy, momentum and angular momentum intrinsically originate from the fundamental properties of time and space, uniformity and isotropy. By way of explanation, the energy conservation law is associated with uniformity of time, while the laws of conservation of momentum and angular momentum with uniformity and isotropy of space respectively. This implies that the conservation laws listed above can be derived from Newton's second law supplemented with the corresponding properties of time and space symmetry. We shall not, however, discuss this problem in more detail.

The laws of conservation of energy, momentum and angular momentum fall into the category of the most fundamental principles of physics, whose significance cannot be overestimated. These laws have become even more significant since it was discovered that they go beyond the scope of mechanics and represent universal laws of nature. In any case, no phenomena have been observed so far which do not obey these laws. They "work" reliably in all quarters: in the field of elementary particles, in outer space, in atomic physics and in solid state physics. They are among the few most general laws underlying contemporary physics.

Having made possible a new approach to treating various mechanical phenomena, the conservation laws turned into a powerful and efficient instrument of research used by physicists. The importance of the conservation principles as a research instrument is due to several reasons.

1. The conservation laws do not depend on either the paths of particles or the nature of acting forces. Consequently, they allow us to draw some general and essential conclusions about the properties of various mechanical processes without resorting to their detailed analysis by means of motion equations. For example, as soon as it turns out that a certain process is in conflict with the conservation laws, one can be sure that such a process is impossible and it is no

use trying to accomplish it.

2. Since the conservation laws do not depend on acting forces, they may be employed even when the forces are not known. In these cases the conservation laws are the only and indispensable instrument of research. This is the present trend in the physics of elementary particles.

3. Even when the forces are known precisely, the conservation laws can help substantially to solve many problems of motion of particles. Although all these problems can be solved with the use of motion equations (and the conservation laws provide no additional information in this case), the utilization of the conservation laws very often allows the solution to be obtained in the most straightforward and elegant fashion, obviating cumbersome and tedious calculations. Therefore, whenever new problems are ventured, the following order of priorities should be established: first, one after another conservation laws are applied and only having made sure that they are inadequate, the solution is sought through the use of motion equations.

We shall begin examining the conservation laws with the energy conservation law, having introduced the notion of energy via the notion of work.

§ 3.2. Work and Power

Work. Let a particle travel along a path *1-2* (Fig. 37) under the action of the force **F**. In the general case the force **F** may vary during the motion, both in magnitude and direction. Let us consider the elementary displacement $d\mathbf{r}$, during which the force **F** can be assumed constant.

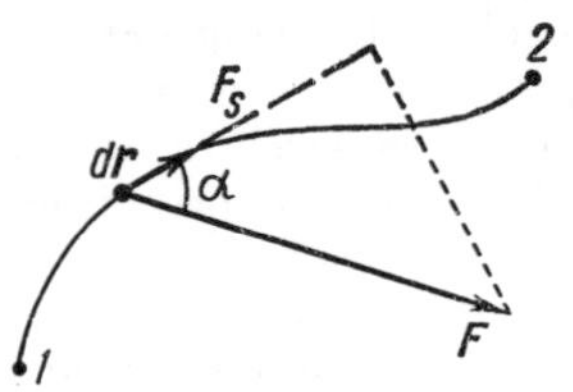

Fig. 37

The action of the force **F** over the displacement $d\mathbf{r}$ is characterized by a quantity equal to the scalar product $\mathbf{F}d\mathbf{r}$ and called the *elementary work* of the force **F** over the displacement $d\mathbf{r}$. It can also be presented in another form:

$$\mathbf{F}d\mathbf{r} = F \cos \alpha ds = F_s ds,$$

where α is the angle between the vectors **F** and $d\mathbf{r}$, $ds = |d\mathbf{r}|$ is the elementary path, and F_s is the projection of the vector **F** on the vector $d\mathbf{r}$ (Fig. 37).

Thus, the elementary work of the force **F** over the displacement $d\mathbf{r}$ is

$$\delta A = \mathbf{F}d\mathbf{r} = F_s ds. \tag{3.1}$$

The quantity δA is algebraic: depending on the angle between the vectors **F** and $d\mathbf{r}$, or on the sign of the projection F_s of the vector **F** on the vector $d\mathbf{r}$, it can be either positive or negative, or, in particular, equal to zero (when $\mathbf{F} \perp d\mathbf{r}$, i.e. $F_s = 0$).

Summing up (integrating) the expression (3.1) over all elementary sections of the path from point *1* to point *2*, we find the work of the force **F** over the given path:

$$\boxed{A = \int_1^2 \mathbf{F}\, d\mathbf{r} = \int_1^2 F_s\, ds.} \tag{3.2}$$

The expression (3.2) can be graphically illustrated. Let us plot F_s as a function of the particle position along the path. Suppose, for example, that this plot has the shape shown in Fig. 38. From this figure the elementary work δA is seen to be numerically equal to the area of the shaded strip, and the work A over the path from point *1* to point *2* is equal to the area of the figure enclosed by the curved line, ordinates *1* and *2*, and the s axis. Here the area of the figure lying over the s axis is taken with the plus sign (it corresponds to positive work) while the area of the figure lying under the s axis is taken with the minus sign (it corresponds to negative work).

Let us consider a few examples involving calculations of work.

The work of the elastic force $\mathbf{F} = -\varkappa\mathbf{r}$, where **r** is the radius vector of the particle A relative to the point O (Fig. 39). Let us displace the particle A experiencing the action of that force along an arbitrary path from point *1* to point *2*. We shall first find the elementary work performed by the force **F** over the elementary displacement $d\mathbf{r}$:

$$\delta A = \mathbf{F}\, d\mathbf{r} = -\varkappa\mathbf{r}\, d\mathbf{r}.$$

The scalar product $\mathbf{r}\,d\mathbf{r} = r\,(d\mathbf{r})_r$, where $(d\mathbf{r})_r$ is the projection of $d\mathbf{r}$ on the vector $\mathbf{r}$. This projection is equal to dr, the increment of the magnitude of the vector $\mathbf{r}$. Therefore $\mathbf{r}\,d\mathbf{r} = r\,dr$ and

$$\delta A = -\varkappa r\,dr = -d\,(\varkappa r^2/2).$$

Now, to calculate the work performed by the given force over the whole path, we should integrate the last expression between point *1* and point *2*:

$$A = -\int_1^2 d\,(\varkappa r^2/2) = \varkappa r_1^2/2 - \varkappa r_2^2/2. \tag{3.3}$$

The work of a gravitational (or Coulomb's) force. Let a stationary point mass (charge) be positioned at the point O

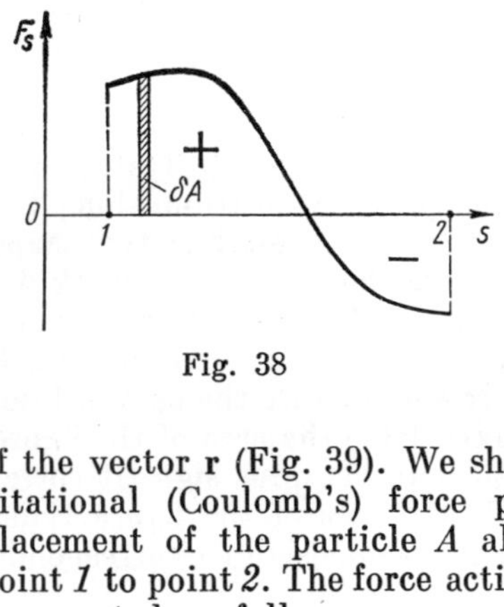

Fig. 38

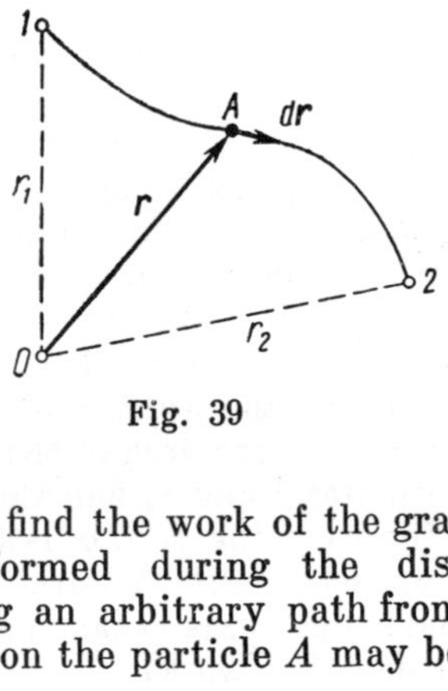

Fig. 39

of the vector $\mathbf{r}$ (Fig. 39). We shall find the work of the gravitational (Coulomb's) force performed during the displacement of the particle A along an arbitrary path from point *1* to point *2*. The force acting on the particle A may be represented as follows:

$$\mathbf{F} = (\alpha/r^3)\,\mathbf{r},$$

where $\alpha = \begin{cases} -\gamma\, m_1 m_2, & \text{the gravitational interaction,} \\ k q_1 q_2, & \text{the Coulomb interaction.} \end{cases}$

Let us first calculate the elementary work performed by this force over the displacement $d\mathbf{r}$:

$$\delta A = \mathbf{F}\,d\mathbf{r} = (\alpha/r^3)\,\mathbf{r}\,d\mathbf{r}.$$

As in the previous case, the scalar product $\mathbf{r}\,d\mathbf{r} = r\,dr$, so that

$$\delta A = \alpha\,dr/r^2 = -d(\alpha/r).$$

The total work performed by this force over the whole path from point *1* to point *2* is

$$A = -\int_1^2 d(\alpha/r) = \alpha/r_1 - \alpha/r_2. \tag{3.4}$$

The work of the uniform force of gravity $\mathbf{F} = m\mathbf{g}$. Let us write this force as $\mathbf{F} = -mg\mathbf{k}$, where $\mathbf{k}$ is the unit vector of the vertical z axis whose positive direction is chosen upward (Fig. 40). The elementary work of gravity over the displacement $d\mathbf{r}$ is

$$\delta A = \mathbf{F}\,d\mathbf{r} = -mg\mathbf{k}d\mathbf{r}.$$

The scalar product $\mathbf{k}\,d\mathbf{r} = (d\mathbf{r})_\mathbf{k}$, where $(d\mathbf{r})_\mathbf{k}$ is the projection of $d\mathbf{r}$ on the unit vector $\mathbf{k}$ and is equal to dz, the z coordinate increment. Therefore, $\mathbf{k}\,d\mathbf{r} = dz$ and

$$\delta A = -mg\,dz = -d(mgz).$$

The total work of this force performed over the whole path from point *1* to point *2* is

$$A = -\int_1^2 d(mgz) = mg(z_1 - z_2). \tag{3.5}$$

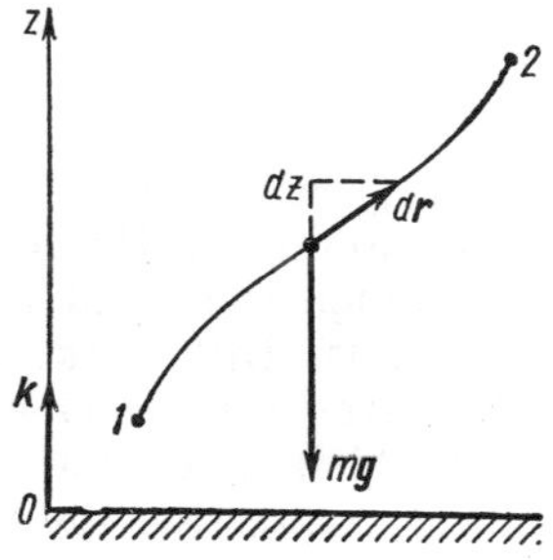

Fig. 40

The forces considered are interesting in that the work performed by them between points *1* and *2* does not depend on the shape of the path and is determined only by their positions (see Eqs. (3.3)-(3.5)). However, this very significant peculiarity of the forces considered is by no means a property of all forces. For example, the friction force does not possess this property: the work performed by this force depends not only on the positions of the initial and final points but also on the shape of the path connecting them.

So far we have discussed the work performed by a single force. But if during its motion a particle experiences several forces whose resultant is $\mathbf{F} = \mathbf{F}_1 + \mathbf{F}_2 + \ldots$, it can be easily shown that the work performed by the resultant force $\mathbf{F}$ over a certain displacement is equal to the algebraic sum of the works performed by all the forces over the same displacement. In fact,

$$A = \int (\mathbf{F}_1 + \mathbf{F}_2 + \ldots)\, d\mathbf{r} = \int \mathbf{F}_1\, d\mathbf{r} + \int \mathbf{F}_2\, d\mathbf{r} + \ldots = A_1 + A_2 + \ldots \,. \tag{3.6}$$

Power. To characterize intensity of the work performed, the quantity called *power* is introduced. Power is defined as the work performed by a force per unit time. If the force $\mathbf{F}$ performs the work $\mathbf{F}\, d\mathbf{r}$ during the time interval dt, the power developed by that force at a given moment of time is equal to $N = \mathbf{F}\, d\mathbf{r}/dt$; taking into account that $d\mathbf{r}/dt = \mathbf{v}$, we obtain

$$\boxed{N = \mathbf{F}\mathbf{v}.} \tag{3.7}$$

Thus, the power developed by the force $\mathbf{F}$ is equal to the scalar product of the force vector by the vector of velocity with which the point moves under the action of the given force. Just like work, power is an algebraic quantity.

Knowing the power of the force $\mathbf{F}$, we can also find the work which that force performs during the time interval t. Indeed, expressing the integrand in formula (3.2) as $\mathbf{F}\, d\mathbf{r} = \mathbf{F}\mathbf{v}\, dt = N\, dt$, we get

$$A = \int_0^t N\, dt.$$

As an example, see Problem 3.1.

Finally, one very essential circumstance should be pointed out. When dealing with work (or power), in each specific case one should indicate *precisely what force* (or forces) performs that work. Otherwise, misinterpretations are, as a rule, inevitable.

§ 3.3. Potential Field of Forces

A field of force is a region of space at whose each point a particle experiences a force varying regularly from point to point, e.g. the Earth's gravitational field, or the field of the resistance forces in a fluid stream. If the force at each point of a field of forces does not vary in the course of time, such a field is referred to as *stationary*. Obviously, a stationary field of forces may turn into a non-stationary field on transition from one reference frame to another. In a stationary field of forces the force is determined only by the position of a particle.

Generally speaking, the work performed by the forces of the field during the displacement of a particle from point *1* to point *2* depends on the path. However, there are some stationary fields of forces in which that work does not depend on the path between points *1* and *2*. This class of fields possesses a number of the most important properties in physics. Now we shall proceed to these properties.

The definition: a stationary field of forces in which the work performed by these forces between any two points does not depend on the shape of the path but only on the positions of these points, is referred to as *potential*, while the forces themselves are called *conservative*.

If this condition is not satisfied, the field of forces is not potential and the forces of the field are called non-conservative. Among them are, for example, friction forces (the work performed by these forces depends on the path in the general case).

An example of two stationary fields of forces, one of which is potential and the other is not, is examined in Problem 3.2.

Let us demonstrate that *in a potential field the work performed by the field forces over any closed path is equal to zero.* In fact, any closed path (Fig. 41) may be arbitrarily subdivided into two parts: *1a2* and *2b1*. Since the field is potential, then by the hypothesis $A_{12}^{(a)} = A_{12}^{(b)}$. On the other hand, it is obvious that $A_{12}^{(b)} = -A_{21}^{(b)}$. Therefore,

$$A_{12}^{(a)} + A_{21}^{(b)} = A_{12}^{(a)} - A_{12}^{(b)} = 0,$$

which was to be proved.

Conversely, if the work performed by the field forces over any closed path is equal to zero, the work of the same forces performed between arbitrary points *1* and *2* does not depend on the path, i.e. the field is potential. To prove this, let us take two arbitrary paths: *1a2* and *1b2* (Fig. 41). We can connect them to make closed path *1a2b1*. The work performed over this closed path is equal to zero by the hypothesis, i.e. $A_{12}^{(a)}+A_{21}^{(b)}=0$. Hence, $A_{12}^{(a)}=-A_{21}^{(b)}$. But $A_{21}^{(b)}=$ $=-A_{12}^{(b)}$ and therefore

$$A_{12}^{(a)}=A_{12}^{(b)}.$$

Thus, when the work of the field forces performed over any closed path is equal to zero, we obtain the necessary and sufficient condition for the work to be independent of the shape of the path, a fact that can be regarded as a distinctive attribute of any potential field of forces.

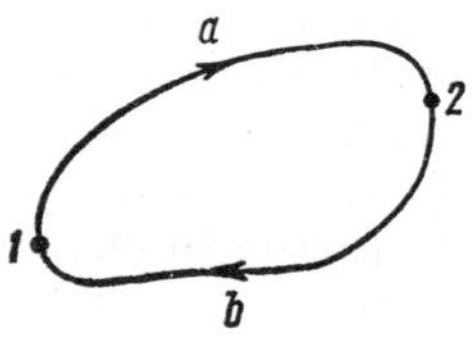

Fig. 41

The field of central forces. Any field of forces is brought about by the action of definite bodies. In that field the particle A experiences a force arising due to the interaction of this particle with the given bodies. Forces depending only on the distance between interacting particles and directed along the straight line connecting them are referred to as *central*. An example of this kind of forces is provided by gravitational, Coulomb's and elastic forces.

The central force which the particle B exerts on the particle A can be presented in general form as follows:

$$\mathbf{F}=f(r)\,\mathbf{e}_r, \tag{3.8}$$

where $f(r)$ is a function depending for the given type of interaction only on r, the distance between the particles; and $\mathbf{e}_r$ is a unit vector defining the direction of the radius vector of the particle A with respect to the particle B (Fig. 42).

Let us demonstrate that *any stationary field of central forces is potential.*

To do this, we shall first find the work performed by the central forces in the case when the field of forces is brought about by one motionless particle B. The elementary work performed by the force (3.8) over the displacement $d\mathbf{r}$ is equal to $\delta A = \mathbf{F}\, d\mathbf{r} = f(r)\, \mathbf{e}_r\, d\mathbf{r}$. Since $\mathbf{e}_r\, d\mathbf{r}$ is equal to dr, the projection of the vector $d\mathbf{r}$ on the vector $\mathbf{e}_r$ or on the corresponding radius vector $\mathbf{r}$ (Fig. 42), then $\delta A = f(r)\, dr$. The work performed by this force over an arbitrary path from point *1* to point *2* is

$$A_{12} = \int_1^2 f(r)\, dr.$$

This expression obviously depends only on the appearance of the function $f(r)$, i.e. on the type of interaction, and on the values of r_1 and r_2, the initial and final distances between the particles A and B. It does not depend on the shape of the path in any way. This means that the given field of forces is potential.

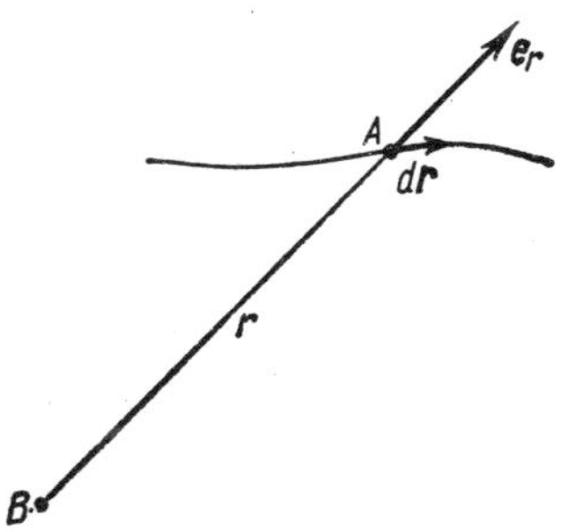

Fig. 42

Let us generalize the result obtained to a stationary field of forces induced by a set of motionless particles exerting on the particle A the forces $\mathbf{F}_1, \mathbf{F}_2, \ldots$, each of which is central. In this case the work performed by the resultant force during the displacement of the particle A from one point to another is equal to the algebraic sum of the works performed by individual forces. But since the work performed by each of these forces does not depend on the shape of the path, the work performed by the resultant force does not depend on it either.

Thus, any stationary field of central forces is indeed potential.

Potential energy of a particle in a field. Due to the fact that the work performed by potential field forces depends only on the initial and final positions of a particle we can introduce the extremely important concept of potential energy.

Suppose we displace a particle from various points P_i of a potential field of forces to a fixed point O. Since the work performed by the field forces does not depend on the shape of the path, it is only the position of the point P that determines this work, provided the point O is fixed. This means that a given work is a certain function of the radius vector **r** of the point P.

Designating this function as $U(\mathbf{r})$, we write

$$A_{PO} = \int_P^O \mathbf{F}\, d\mathbf{r} = U(\mathbf{r}). \tag{3.9}$$

The function $U(\mathbf{r})$ is referred to as the *potential energy* of the particle in a given field.

Now let us find the work performed by the field forces during the displacement of the particle from point *1* to point *2* (Fig. 43). The work being independent of the shape of the path, we can choose one passing through the point O. Then the work performed over the path *1O2* can be represented in the form

$$A_{12} = A_{1O} + A_{O2} = A_{1O} - A_{2O}$$

or, taking into account Eq. (3.9),

$$\boxed{A_{12} = \int_1^2 \mathbf{F}\, d\mathbf{r} = U_1 - U_2.} \tag{3.10}$$

The expression on the right-hand side of this equation is the *diminution** of the potential energy, i.e. the difference

* The variation of some quantity X can be characterized either by its *increment*, or its *diminution*. The increment of the X quantity is the difference between its final (X_2) and initial (X_1) values:

$$\text{the increment } \Delta X = X_2 - X_1.$$

The diminution of the quantity X is taken to be the difference between its initial (X_1) and final (X_2) values:

$$\text{the diminution } X_1 - X_2 = -\Delta X,$$

i.e. the diminution of the quantity X is equal to its increment in magnitude but is opposite in sign.

The diminution and increment are *algebraic* quantities: if $X_2 > X_1$, then the increment is positive while the diminution is negative, and vice versa.

between the values of the potential energy at the initial and final points of the path.

Thus, *the work of the field forces performed over the path 1-2 is equal to the decrease in the potential energy of the particle in a given field.*

Obviously, any potential energy value can be chosen in advance and assigned to a particle positioned at the point O of the field. Therefore, measuring the work makes it possible to determine only the difference of the potential energies at two points but not the absolute value of the potential energy. However, as soon as the potential energy at some point is specified, its values are uniquely determined at all other points of the field by means of Eq. (3.10).

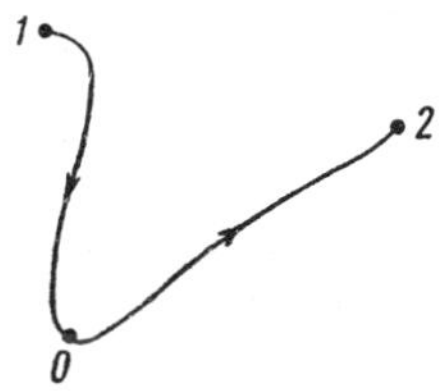

Fig. 43

Eq. (3.10) makes it possible to find the expression $U(\mathbf{r})$ for any potential field of forces. It is sufficient to calculate the work performed by the field forces over any path between two points and to represent it as the diminution of a certain function which is the potential energy $U(\mathbf{r})$.

This is exactly how the work was calculated in the cases of the fields of elastic and gravitational (Coulomb's) forces, as well as in the uniform field of gravity (see Eqs. (3.3)-(3.5)). It is immediately seen from these formulae that the potential energy of a particle in such fields of forces takes the following form:

(1) in the field of elastic forces

$$U(r) = \varkappa r^2/2; \tag{3.11}$$

(2) in the field of a point mass (charge)

$$U(r) = \alpha/r, \tag{3.12}$$

where $\alpha = \begin{cases} -\gamma m_1 m_2, & \text{gravitational interaction,} \\ kq_1q_2, & \text{Coulomb's interaction;} \end{cases}$

(3) in the uniform field of gravity

$$U(z) = mgz. \tag{3.13}$$

It should be pointed out once again that the potential energy U is a function determined with an accuracy of a certain arbitrary addendum. This vagueness, however, is quite immaterial as all equations deal only with the difference of the values of U at two positions of a particle. Therefore, the arbitrary addendum, being equal at all points of the field, gets eliminated. Accordingly, it is usually omitted as it is done in the three previous expressions.

There is another important point. Potential energy should not be assigned to a particle but to a system consisting of this particle interacting with the bodies generating a field of force. For a given type of interaction the potential energy of interaction of a particle with given bodies depends only on the position of the particle relative to these bodies.

Potential energy and force of a field. The interaction of a particle with surrounding bodies can be described in two ways: by means of forces or through the use of the notion of potential energy. In classical mechanics both ways are extensively used. The first approach, however, is more general because of its applicability to forces in the case of which the potential energy is impossible to introduce (e.g. friction forces). As to the second method, it can be utilized only in the case of conservative forces.

Our objective is to establish the relationship between potential energy and the force of the field, or putting it more precisely, to define the field of forces $\mathbf{F}(\mathbf{r})$ from a given potential energy $U(\mathbf{r})$ as a function of a position of a particle in the field.

We have learned by now that the work performed by field forces during the displacement of a particle from one point of a potential field to another may be described as the *diminution* of the potential energy of the particle, that is, $A_{12} = U_1 - U_2 = -\Delta U$. The same can be said about the elementary displacement $d\mathbf{r}$ as well: $\delta A = -dU$, or

$$\mathbf{F}\, d\mathbf{r} = -dU. \tag{3.14}$$

Recalling (see Eq. (3.1)) that $\mathbf{F}\, d\mathbf{r} = F_s\, ds$, where $ds = |\, d\mathbf{r}\,|$ is the elementary path and F_s is the projection of the vector $\mathbf{F}$ on the displacement $d\mathbf{r}$, we shall rewrite Eq. (3.14) as

$$F_s\, ds = -dU,$$

where $-dU$ is the diminution of the potential energy in the $d\mathbf{r}$ *displacement direction*. Hence,

$$F_s = -\partial U/\partial s, \tag{3.15}$$

i.e. the projection of the field force, the vector **F**, at a given point in the direction of the displacement $d\mathbf{r}$ equals the derivative of the potential energy U with respect to a given direction, taken with the opposite sign. The designation of a *partial* derivative $\partial/\partial s$ emphasizes the fact of deriving with respect to a *definite* direction.

The displacement $d\mathbf{r}$ can be accomplished along any direction and, specifically, along the x, y, z coordinate axes. For example, if the displacement $d\mathbf{r}$ is parallel to the x axis, it may be described as $d\mathbf{r} = \mathbf{i}\, dx$, where **i** is the unit vector of the x axis and dx is the x coordinate increment. Then the work performed by the force **F** over the displacement $d\mathbf{r}$ parallel to the x axis is

$$\mathbf{F}\, d\mathbf{r} = \mathbf{F}\mathbf{i}\, dx = F_x\, dx,$$

where F_x is the projection of the vector **F** on the unit vector **i** (but not on the $d\mathbf{r}$ displacement as in the case of F_s).

Substituting the last expression into Eq. (3.14), we get

$$F_x = -\partial U/\partial x,$$

where the partial derivative symbol implies that in the process of differentiation $U(x, y, z)$ should be considered as a function of only one variable, x, while all other variables are assumed constant. It is obvious that the equations for the F_y and F_z projections are similar to that for F_x.

So, having reversed the sign of the partial derivatives of the function U with respect to x, y, z, we obtain the projections F_x, F_y and F_z of the vector **F** on the unit vectors **i**, **j** and **k**. Hence, one can readily find the vector itself: $\mathbf{F} = F_x\mathbf{i} + F_y\mathbf{j} + F_z\mathbf{k}$, or

$$\mathbf{F} = -\left(\frac{\partial U}{\partial x}\mathbf{i} + \frac{\partial U}{\partial y}\mathbf{j} + \frac{\partial U}{\partial z}\mathbf{k}\right).$$

The quantity in parentheses is referred to as the *scalar gradient of the function U* and is denoted by grad U or ∇U. We shall use the second, more convenient, designation where

∇ ("nabla") signifies the symbolic vector or operator

$$\nabla = \mathbf{i}\frac{\partial}{\partial x} + \mathbf{j}\frac{\partial}{\partial y} + \mathbf{k}\frac{\partial}{\partial z}.$$

Therefore ∇U may be formally regarded as the product of a symbolic vector ∇ by a scalar U.

Consequently, the relationship between the force of a field and the potential energy, expressed as a function of coordinates, can be written in the following compact form:

$$\boxed{\mathbf{F} = -\nabla U,} \tag{3.16}$$

i.e. *the field force* **F** *is equal to the potential energy gradient, taken with the minus sign*, for a particle at a given point of the field. Put simply, the field force **F** is equal to the *anti-gradient* of potential energy. The last equation permits the field of forces **F** (**r**) to be derived from the function U (**r**).

Example. The potential energy of a particle in a certain field has the following form:

(a) $U(x, y) = -\alpha xy$, where α is a constant;

(b) $U(\mathbf{r}) = \mathbf{ar}$, where **a** is a constant vector and **r** is the radius vector of a point of the field.

Let us find the field of force corresponding to each of these cases:

(a) $\mathbf{F} = -\left(\frac{\partial U}{\partial x}\mathbf{i} + \frac{\partial U}{\partial y}\mathbf{j}\right) = \alpha(y\mathbf{i} + x\mathbf{j});$

(b) first, let us transform the function U to the following form: $U = a_x x + a_y y + a_z z$; then

$$\mathbf{F} = -\left(\frac{\partial U}{\partial x}\mathbf{i} + \frac{\partial U}{\partial y}\mathbf{j} + \frac{\partial U}{\partial z}\mathbf{k}\right) = -(a_x\mathbf{i} + a_y\mathbf{j} + a_z\mathbf{k}) = -\mathbf{a}.$$

The meaning of a gradient becomes more obvious and descriptive as soon as we introduce the concept of an *equipotential surface* at all of whose points the potential energy U has the same magnitude. It is clear that each value of U has a corresponding equipotential surface.

It follows from Eq. (3.15) that the projection of the vector **F** on any direction tangential to the equipotential surface at a given point is equal to zero. This means that the vector **F** is normal to the equipotential surface at a given point. Next, let us consider the displacement ∂s in the direction of decreasing U values; then $\partial U < 0$ and in accordance with

Eq. (3.15) $F_s > 0$, i.e. the vector **F** is oriented in the direction of decreasing U values. As **F** is directed oppositely to the vector ∇U, we may conclude that *the gradient of U is a vector oriented along a normal to an equipotential surface in the direction of increasing values of potential energy U*.

Fig. 44 illustrating a two-dimensional case clarifies what was said above. It shows a system of equipotentials ($U_1 < U_2 < U_3 < U_4$), a gradient of the potential energy ∇U and the corresponding vector of the force **F** at the point A of the field. It pays to consider how these two vectors are directed, for example, at the point B of the given field.

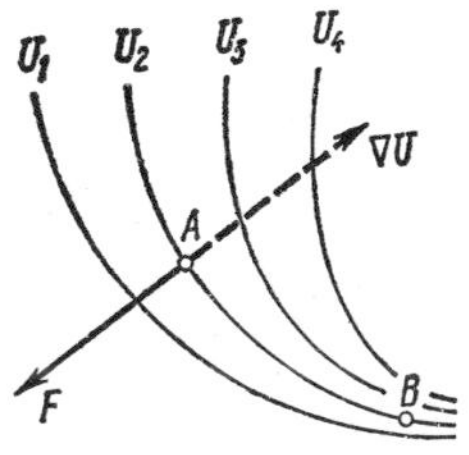

Fig. 44

In conclusion it should be pointed out that a gradient can be calculated not only from the function U but from any other scalar function of coordinates. The concept of a gradient permeates various divisions of physics.

Concept of a field. Experience shows that in the case of gravitational and electrostatic interactions the force **F** that surrounding bodies (the system B) exert on a particle A is proportional to the mass (or the charge) of that particle. In other words, the force **F** may be represented as the product of two quantities:

$$\mathbf{F} = m\mathbf{G}, \tag{3.17}$$

where m is the mass (or the charge) of the particle A, and **G** is a certain vector depending both on the position of the particle A and on the properties of surrounding bodies, that is, the system B.

This opens up the possibility for another physical interpretation of interaction which is based on the concept of a field. In this case the system B is assumed to induce a spatial *field* characterized by the vector **G** (**r**). Expressed otherwise: the system B, the origin of a field, is assumed to provide such conditions (the vector **G**) at all points of space that a particle positioned at any point experiences the force (3.17). Moreover, the field (the vector **G**) is supposed to exist irrespective of whether the particle A is actually in the field or not*.

The vector **G** is referred to as the *field strength.*

One of the most important properties of a field is that the field induced by several sources is equal to the sum of the fields induced by all of them. Putting it more precisely, the strength **G** of the resultant field at an arbitrary point

$$\mathbf{G} = \sum \mathbf{G}_i, \tag{3.18}$$

where $\mathbf{G}_i$ is the field strength of the ith source at the same point. This formula expresses the so-called *principle of superposition* of fields.

Now let us direct our attention to the potential energy of a particle. In accordance with Eq. (3.17) we can rewrite Eq. (3.14) as $m\mathbf{G}\,d\mathbf{r} = -dU$. Dividing both sides by m and denoting $U/m = \varphi$, we get

$$\mathbf{G}\,d\mathbf{r} = -d\varphi \tag{3.19}$$

or

$$\int_1^2 \mathbf{G}\,d\mathbf{r} = \varphi_1 - \varphi_2. \tag{3.20}$$

The function $\varphi(\mathbf{r})$ is called the *field potential* at the point whose radius vector is equal to **r**.

Eq. (3.20) allows the potential of any gravitational or electrostatic field to be found. To do this, it is sufficient to calculate the integral $\int \mathbf{G}\,d\mathbf{r}$ over an arbitrary path between points *1* and *2* and then to present the expression obtained as a diminution of a certain function, the potential $\varphi(\mathbf{r})$. For instance, the potentials of the gravitational field of a point mass m and of the Coulomb field of a point charge q are determined, in accordance with Eq. (3.12), as follows:

$$\varphi_{gr} = -\gamma m/r, \quad \varphi_{Coul} = kq/r. \tag{3.21}$$

* While we confine ourselves to statics, the concept of a field may be treated as a mere formality simplifying the description of phenomena. However, when we pass to variable fields, the concept of a field acquires a profound physical meaning: *a field is a physical reality.*

Note that the potential φ, as well as the potential energy, can be determined only with an accuracy of some addendum, whose magnitude is of no importance. Accordingly, it is usually omitted.

Thus, a field can be described either in vector form $\mathbf{G}(\mathbf{r})$ or in scalar form $\varphi(\mathbf{r})$. Both methods are adequate. However, for all practical purposes the second method of describing a field (by means of the potential φ) turns out to be far more convenient in most cases. Here is why this is so.

1. If $\varphi(\mathbf{r})$ is known, the potential energy U and the work A performed by the field forces can be immediately obtained:

$$U = m\varphi, \quad A_{12} = m(\varphi_1 - \varphi_2). \tag{3.22}$$

2. Instead of the three components of the vectorial function $\mathbf{G}(\mathbf{r})$ it is simpler to specify the scalar function $\varphi(\mathbf{r})$.

3. When a field is produced by many sources, the potential φ is more easily calculated than the vector $\mathbf{G}$: potentials are scalars, so that they can be summed up disregarding the force directions. In fact, in accordance with Eqs. (3.18) and (3.19) $\mathbf{G}\,d\mathbf{r} = \sum (\mathbf{G}_i\, d\mathbf{r}) = -\sum d\varphi_i = -d\sum \varphi_i = -d\varphi$ i.e.

$$\varphi(\mathbf{r}) = \sum \varphi_i(\mathbf{r}), \tag{3.23}$$

where φ_i is the potential produced by the ith particle at a given point of the field.

4. And finally, when the function $\varphi(\mathbf{r})$ is known, one can readily obtain the field $\mathbf{G}(\mathbf{r})$ as the antigradient of the potential φ:

$$\mathbf{G} = -\nabla\varphi. \tag{3.24}$$

This formula follows directly from Eq. (3.16).

In conclusion let us examine an example involving the determination of the field potential of centrifugal forces of inertia.

Example. Let us find the field strength $\mathbf{G}$ and the potential φ_{cf} of centrifugal forces of inertia in a reference frame rotating with a constant angular velocity about a stationary axis.

The field strength $\mathbf{G} = \mathbf{F}_{cf}/m = \omega^2 \boldsymbol{\rho}$, where $\boldsymbol{\rho}$ is the radius vector of a point of the field relative to the rotation axis.

Now, utilizing Eq. (3.20), we integrate G over a path from point *1* to point *2*:

$$\int_1^2 \mathbf{G}\, d\mathbf{r} = \omega^2 \int_1^2 \boldsymbol{\rho}\, d\mathbf{r} = \omega^2 \int_1^2 \rho\, d\rho = \omega^2 (\rho_2^2 - \rho_1^2)/2.$$

It is evident that this integral does not depend on the shape of the path between points *1* and *2* and is defined by the positions of the given points. This means that the considered field of forces is potential.

Comparing the result obtained with Eq. (3.20), we get

$$\varphi_{cf} = -\omega^2\rho^2/2. \tag{3.25}$$

Fig. 45 shows φ_{cf} vs the distance ρ to the rotation axis. For the sake of comparison also shown is the potential $\varphi_{gr}(\rho)$ of the gravitational field produced by a point mass placed at the point $\rho = 0$.

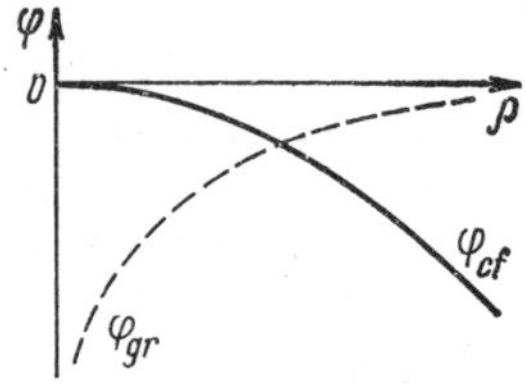

Fig. 45

§ 3.4. Mechanical Energy of a Particle in a Field

Kinetic energy. Suppose a particle of mass m moves under the action of a certain force $\mathbf{F}$, which in the general case may be a resultant of several forces. Let us determine the elementary work performed by this force over an elementary displacement $d\mathbf{r}$. Taking into account that $\mathbf{F} = m\, d\mathbf{v}/dt$ and $d\mathbf{r} = \mathbf{v}\, dt$, we may write

$$\delta A = \mathbf{F}\, d\mathbf{r} = m\mathbf{v}\, d\mathbf{v}.$$

The scalar product $\mathbf{v}\, d\mathbf{v} = v\, (d\mathbf{v})_{\mathbf{v}}$, where $(d\mathbf{v})_{\mathbf{v}}$ is the projection of the vector $d\mathbf{v}$ on the direction of the vector $\mathbf{v}$. This projection is equal to dv, the increment of the velocity vector magnitude. Consequently, $\mathbf{v}\, d\mathbf{v} = v\, dv$ and the elementary work

$$\delta A = mv\, dv = d\,(mv^2/2).$$

It is seen from this that the work of the resultant force $\mathbf{F}$ contributes to an increase in a certain quantity (enclosed in parentheses) referred to as *kinetic energy*:

$$\boxed{T = mv^2/2.} \qquad (3.26)$$

Thus, the increment of the kinetic energy of a particle during its elementary displacement is equal to

$$dT = \delta A, \qquad (3.27)$$

and during the finite displacement from point *1* to point *2*

$$\boxed{T_2 - T_1 = A_{12},} \qquad (3.28)$$

i.e. *the increment of the kinetic energy gained by a particle over a certain displacement is equal to the algebraic sum of the works performed by all forces* acting on the particle over the same displacement. If $A_{12} > 0$, then $T_2 > T_1$, i.e. the kinetic energy of a particle grows; but if $A_{12} < 0$, the kinetic energy decreases.

Eq. (3.27) may be written in another form, having divided its both sides by the corresponding time interval dt:

$$\boxed{dT/dt = \mathbf{F}\mathbf{v} = N.} \qquad (3.29)$$

This means that the derivative of the kinetic energy of a particle with respect to time is equal to the power N developed by the resultant force $\mathbf{F}$ acting on the particle.

Eqs. (3.28) and (3.29) hold true both in inertial and in non-inertial reference frames. In the latter frames, apart from the forces exerted on the considered particle by other bodies (interaction forces), we must also take into account inertial forces. That is why in these equations work (power) should be conceived as an algebraic sum of the works (powers) performed both by interaction forces and by inertial forces.

Total mechanical energy of a particle. In accordance with Eq. (3.27) the increment of the kinetic energy of a particle is equal to the elementary work performed by the resultant $\mathbf{F}$ of all forces acting on the particle. What kind of forces are they? If the particle is located in a potential field, this field exerts a conservative force $\mathbf{F}_c$ on it. Besides, the particle may experience other forces of different origin. We shall call them *outside forces* $\mathbf{F}_{out}$.

Thus, the resultant **F** of all forces acting on the particle may be presented in the form $\mathbf{F} = \mathbf{F}_c + \mathbf{F}_{out}$. The work performed by all these forces results in an increment of the kinetic energy of the particle:

$$dT = \delta A_c + \delta A_{out}.$$

In accordance with Eq. (3.14) the work performed by the field forces is equal to the decrease in the potential energy of the particle, i.e. $\delta A_c = -dU$. Substituting this expression into the previous one and transposing the term dU to the left-hand side, we obtain

$$dT + dU = d(T+U) = \delta A_{out}.$$

It is seen from this that the work performed by the outside forces results in an increment of the quantity $T + U$. This quantity, the sum of the kinetic and potential energies, is referred to as *the total mechanical energy of a particle in a field*:

$$\boxed{E = T + U.} \tag{3.30}$$

Note that the total mechanical energy E, just as the potential energy, is defined with an accuracy up to an arbitrary constant.

So, the increment of the total mechanical energy acquired by a particle over an elementary displacement is equal to

$$dE = \delta A_{out} \tag{3.31}$$

and over the finite displacement from point *1* to point *2*

$$\boxed{E_2 - E_1 = A_{out},} \tag{3.32}$$

i.e. *the increment of the total mechanical energy acquired by a particle over a certain path is equal to the algebraic sum of works performed by all outside forces* acting on the particle over the same path. When $A_{out} > 0$, the total mechanical energy of a particle increases, and when $A_{out} < 0$, it decreases.

Example. From a cliff rising to the height h over a lake surface a stone of mass m is thrown with the velocity v_0. Find the work performed by air resistance forces if the stone falls on the lake surface with the velocity v.

When the motion of the body is considered in the Earth's gravitational field, the air resistance forces should be treated as outside ones, and in accordance with Eq. (3.32) the sought work $A_{res} = E_2 - E_1 = mv^2/2 - (mv_0^2/2 + mgh)$, or

$$A_{res} = m(v^2 - v_0^2)/2 - mgh.$$

The obtained value may turn out to be not only negative but positive as well (this depends, for example, on the character of wind during the fall of the body).

Eq. (3.31) may be presented in another form if its both sides are divided by the corresponding time interval dt. Then

$$\boxed{dE/dt = \mathbf{F}_{out}\mathbf{v}.} \qquad (3.33)$$

This implies that the derivative of the total mechanical energy of a particle with respect to time is equal to the power developed by the resultant of all outside forces acting on the particle.

We have thus established that the total mechanical energy of a particle can change only due to outside forces. Hence, the *law of conservation* of the total mechanical energy of a particle in an external field follows directly:

when outside forces are absent or such that the algebraic sum of powers developed by them during the time interval considered is equal to zero, the total mechanical energy of a particle remains constant during that interval. In other words,

$$E = T + U = \text{const},$$

or

$$\boxed{mv^2/2 + U(\mathbf{r}) = \text{const}.} \qquad (3.34)$$

Even when written in such a simple form this conservation law permits some significant solutions to be obtained quite easily without resorting to equations of motion associated with cumbersome and tedious calculations. That is why the conservation laws prove to be a very efficient instrument of research.

The following example illustrates the capabilities and advantages provided by the application of the conservation law (3.34).

Example. Suppose a particle moves in the unidimensional potential field $U(x)$ shown in Fig. 46. In the absence of outside forces the total mechanical energy of a particle in this field, i.e. E, does not vary during the motion, and we can easily solve, for example, the following problems.

1. To find $v(x)$, the velocity of the particle as a function of its coordinate, without solving the fundamental equation of dynamics. In accordance with Eq. (3.34), this can be done provided the specific appearance of the *potential curve* $U(x)$ and the total energy value E (the right-hand side of that equation) are precisely known.

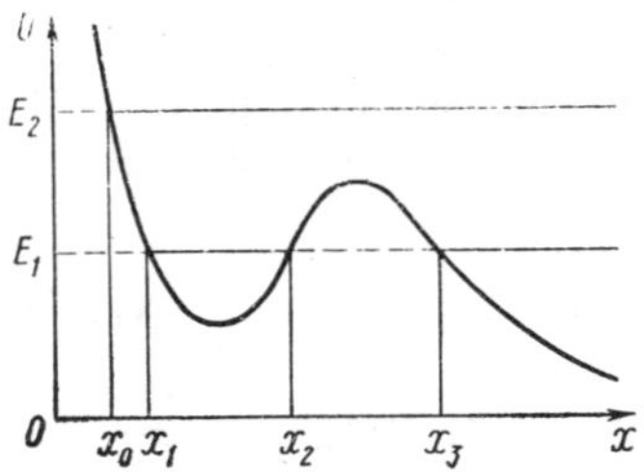

Fig. 46

2. To establish the *variation range* of the x coordinate of the particle within which it can be located when the total energy value E is fixed. It is clear that the particle cannot get in the region where $U > E$ since the potential energy U of the particle cannot exceed its total energy. From this it immediately follows that when $E = E_1$ (Fig. 46), the particle can move either in the region confined by the x_1 and x_2 coordinates (oscillation) or to the right of the x_3 coordinate. The particle cannot however pass from the first region to the second one (or vice versa) due to the *potential barrier* dividing these regions. Note that the particle moving in a confined region of a field is referred to as being locked in a *potential well* (in our case, between x_1 and x_2).

The particle behaves differently when $E = E_2$ (Fig. 46): the whole region lying to the right of x_0 becomes accessible to it. If at an initial moment the particle was located at the point x_0, it travels to the right. The reader is advised to find how the kinetic energy of the particle varies depending on its coordinate x.

§ 3.5. The Energy Conservation Law for a System

Until now we confined ourselves to treating the behaviour of a *single* particle in terms of its energy. Now we shall pass over to a *system of particles*. That may be any object, gas, any device, the solar system, etc.

In the general case particles of a system interact both with one another and with other bodies which do not belong to that system. The system of particles experiencing negligible (or no) action from external bodies is called *closed* (or isolated). The concept of a closed system is the natural generalization of the concept of an isolated mass point and plays a significant part in physics.

Potential energy of a system. Let us consider a closed system whose particles interact only through central forces, i.e. the forces depending (for a given type of interaction) only on a distance between the particles and directed along the straight line that connects them.

We shall show that in any reference frame the work performed by all these forces during the transition of the system of particles from one position to another may be represented as a decrease of a certain function depending (for a given type of interaction) only on the configuration of the system, that is, on the relative positions of its particles. We shall call this function an *internal* potential energy of the system (in distinction to a potential energy characterizing the interaction of a given system with other bodies).

First we shall examine a system consisting of two particles. Let us calculate the elementary work performed by the force of interaction between the particles. Suppose that in an arbitrary reference frame the positions of the particles at a certain moment of time are defined by the radius vectors $\mathbf{r}_1$ and $\mathbf{r}_2$. If during the time dt the particles shift through $d\mathbf{r}_1$ and $d\mathbf{r}_2$ respectively, the work performed by the interaction forces $\mathbf{F}_{12}$ and $\mathbf{F}_{21}$ is equal to

$$\delta A_{1,2} = \mathbf{F}_{12}\, d\mathbf{r}_1 + \mathbf{F}_{21}\, d\mathbf{r}_2.$$

Taking into account that in accordance with Newton's third law $\mathbf{F}_{21} = -\mathbf{F}_{12}$, the previous expression may be rewritten in the following form:

$$\delta A_{1,2} = \mathbf{F}_{12}\,(d\mathbf{r}_1 - d\mathbf{r}_2).$$

Let us introduce the vector $\mathbf{r}_{12} = \mathbf{r}_1 - \mathbf{r}_2$ describing the position of the first particle with respect to the second particle. Then $d\mathbf{r}_{12} = d\mathbf{r}_1 - d\mathbf{r}_2$, substituted into the expression for the work, it yields

$$\delta A_{1,2} = \mathbf{F}_{12}\, d\mathbf{r}_{12}.$$

$\mathbf{F}_{12}$ is a central force, and therefore the work performed by this force is equal to the decrease in the potential energy of interaction of a given pair of particles, i.e.

$$\delta A_{1,2} = -dU_{12}.$$

Since the function U_{12} depends only on the distance r_{12} between the particles, the work $\delta A_{1,2}$ obviously is not affected by the choice of a reference frame.

Now let us direct our attention to a system comprising three particles (in this case the results obtained can readily be generalized for a system consisting of an arbitrary number of particles). The elementary work performed by all interaction forces during elementary displacements of all particles may be presented as the sum of the elementary works performed by all three pairs of interactions, i.e. $\delta A = \delta A_{1,2} + \delta A_{1,3} + \delta A_{2,3}$. It was shown, however, that for each pair of interactions $\delta A_{i,k} = -dU_{ik}$, and therefore

$$\delta A = -d(U_{12} + U_{13} + U_{23}) = -dU,$$

where the function U is the *internal potential energy* of the given system of particles:

$$U = U_{12} + U_{13} + U_{23}.$$

Since each term of this sum depends on the distance between the corresponding particles, the internal potential energy U of the given system obviously depends on the relative positions of particles (at the same moment of time), or in other words, on the configuration of the system.

Naturally, similar arguments hold true for a system comprising any number of particles. Consequently, it can be claimed that *every configuration of an arbitrary system of particles is characterized by its internal potential energy U, and the work performed by all internal central forces, while that configuration varies, is equal to the decrease of the internal potential energy of the system, i.e.*

$$\delta A = -dU; \tag{3.35}$$

during the finite displacement of all particles of the system

$$A = U_1 - U_2, \tag{3.36}$$

where U_1 and U_2 are the values of the potential energy of the system in the initial and final states.

The internal potential energy U of the system is a *non-additive* quantity, that is, in the general case it is not equal to the sum of the internal potential energies of the constituent parts. The potential energy of interaction U_{ia} between the individual parts of the system should be also taken into account:

$$U = \sum U_n + U_{ia}, \tag{3.37}$$

where U_n is the internal potential energy of the nth part of the system.

It should be also borne in mind that the internal potential energy of a system, just as the potential energy of interaction for each pair of particles, is defined with an accuracy up to an arbitrary constant which is quite insignificant, though.

In conclusion, we shall quote some useful formulae for calculation of the internal potential energy of a system. First of all, we shall demonstrate that that energy may be represented as

$$\boxed{U = \frac{1}{2} \sum U_i,} \tag{3.38}$$

where U_i is the potential energy of interaction of the ith particle with all remaining particles of the system. Here the summation is performed over all particles of the system.

First, let us make sure that this formula is valid for a system of three particles. It was shown earlier that the internal potential energy of the given system is $U = U_{12} + U_{13} + U_{23}$. This sum can be transformed as follows. Let us depict each term U_{ik} in a symmetrical form: $U_{ik} = (U_{ik} + U_{ki})/2$, for it is obvious that $U_{ik} = U_{ki}$. Then

$$U = \frac{1}{2}(U_{12} + U_{21} + U_{13} + U_{31} + U_{23} + U_{32}).$$

Grouping together the terms with identical first subindex, we get

$$U = \frac{1}{2}[(U_{12} + U_{13}) + (U_{21} + U_{23}) + (U_{31} + U_{32})].$$

Each sum in parentheses represents the potential energy U_i of interaction of the ith particle with the remaining two.

Consequently, the last expression can be rewritten as

$$U=\frac{1}{2}(U_1+U_2+U_3)=\frac{1}{2}\sum_{i=1}^{3}U_i,$$

identically with Eq. (3.38).

The result obtained can be evidently applied to an arbitrary system since these arguments by no means depend on the number of particles constituting the system.

Making use of the concept of potential, Eq. (3.38) can be transformed in the case of a system with the gravitational or Coulomb interaction between particles. Replacing the potential energy of the ith particle in Eq. (3.38) by $U_i = m_i\varphi_i$, where m_i is the mass (charge) of the ith particle and φ_i is the potential produced by all remaining particles of the system at the site of the ith particle, we obtain

$$U=\frac{1}{2}\sum m_i\varphi_i. \tag{3.39}$$

When the masses (charges) are continuously distributed throughout the system, summation is reduced to integration:

$$U=\frac{1}{2}\int\varphi\,dm=\frac{1}{2}\int\varphi\rho\,dV, \tag{3.40}$$

where ρ is the volume density of the mass (charge), and dV is the volume element. Here integration is performed over the whole volume occupied by the masses (charges).

The application of the last formula may be illustrated by Problem 3.11 in which the internal potential energy of gravitational interaction of masses distributed over the surface and volume of a sphere is calculated.

Kinetic energy of a system. Let us examine an arbitrary system of particles in a certain reference frame. Suppose the ith particle of the system has the kinetic energy T_i at a given moment. In accordance with Eq. (3.26) the increment of the kinetic energy of each particle is equal to the work performed by all forces acting on this particle: $dT_i = \delta A_i$. Let us find the elementary work performed by all forces acting on all particles of the system:

$$\delta A=\sum\delta A_i=\sum dT_i=d\sum T_i=dT,$$

where $T = \sum T_i$ is the total kinetic energy of the system. Note that the kinetic energy of a system is an *additive* quantity: it is equal to the sum of the kinetic energies of individual parts of a system irrespective of whether they interact with one another or not.

So, *the increment of the kinetic energy of a system is equal to the work performed by* ***all*** *the forces acting on* ***all*** *the particles of that system.* During an elementary displacement of all particles

$$dT = \delta A, \tag{3.41}$$

and in the case of a finite displacement

$$\boxed{T_2 - T_1 = A.} \tag{3.42}$$

Eq. (3.41) may be represented in another form, having divided its both sides by the corresponding time interval dt. Making allowance for $\delta A_i = \mathbf{F}_i \mathbf{v}_i \, dt$, we obtain

$$\boxed{dT/dt = \sum \mathbf{F}_i \mathbf{v}_i,} \tag{3.43}$$

i.e. *the time derivative of the kinetic energy of a system is equal to the cumulative power of* ***all*** *the forces acting on* ***all*** *the particles of the system.*

Eqs. (3.41)-(3.43) hold true both in inertial and in non-inertial reference frames. It should be recognized that in non-inertial frames the work performed by interaction forces is to be supplemented by that of inertial forces.

Classification of forces. The particles of the considered system are known to be able to interact both with one another and with bodies outside the given system. Accordingly, the forces of interaction between the particles of the system are referred to as *internal* while the forces caused by the action of other bodies outside the given system are called *external.* In a non-inertial reference frame the latter forces include also inertial ones.

Furthermore, all forces are subdivided into *potential* and *non-potential* ones. The forces are referred to as potential if, for a given type of interaction, they depend only on the configuration of a mechanical system. The work performed by these forces was shown to be equal to the decrease of the potential energy of the system.

To non-potential forces we refer the so-called *dissipative* forces, the friction and resistance forces. The significant feature of these forces is that the total work performed by *internal* dissipative forces of the considered system is negative in any reference frame. Let us demonstrate that.

Any dissipative force may be represented in the form

$$\mathbf{F} = -k(v)\,\mathbf{v}, \tag{3.44}$$

where $\mathbf{v}$ is the velocity of a given body relative to another body (or medium) with which it interacts and $k(v)$ is the positive coefficient depending in the general case on the velocity v. The force $\mathbf{F}$ is always directed oppositely to the vector $\mathbf{v}$. Depending on the choice of a reference frame the work performed by that force can be either positive or negative. However, *the total work performed by all internal dissipative forces is always a negative quantity.*

To prove this, we should point out first of all that the internal dissipative forces in a given system appear in pairs. In accordance with Newton's third law the forces of each pair are equal in magnitude and opposite in direction. Let us find the elementary work performed by an arbitrary pair of dissipative forces of interaction between bodies *1* and *2* in the reference frame where the velocities of these bodies at a given moment are equal to $\mathbf{v}_1$ and $\mathbf{v}_2$:

$$\delta A^{dis} = \mathbf{F}_{12}\mathbf{v}_1\,dt + \mathbf{F}_{21}\mathbf{v}_2\,dt.$$

Making allowance for $\mathbf{F}_{21} = -\mathbf{F}_{12}$, $\mathbf{v}_1 - \mathbf{v}_2 = \mathbf{v}$ (the velocity of body *1* relative to body *2*) and $\mathbf{F}_{12} = -k(v)\,\mathbf{v}$, we may transform the expression for work as follows:

$$\delta A^{dis} = \mathbf{F}_{12}(\mathbf{v}_1 - \mathbf{v}_2)\,dt = -k(v)\,\mathbf{v}\mathbf{v}\,dt = -k(v)\,v^2\,dt.$$

It is seen from this that the work performed by an arbitrary pair of internal dissipative forces of interaction is always negative, and hence the total work performed by all pairs of internal dissipative forces is always negative. Consequently

$$\boxed{A_{int}^{dis} < 0.} \tag{3.45}$$

Energy conservation law. It was shown above that the increment of the kinetic energy of a system is equal to the work performed by *all* forces acting on *all* particles of the system. Classifying these forces into external and internal ones, and the internal forces, in their turn, into potential and dissipative ones, we write the foregoing statement in the following form:

$$dT = \delta A_{ext} + \delta A_{int} = \delta A_{ext} + \delta A_{int}^{pot} + \delta A_{int}^{dis}.$$

Then we take into account that the work performed by internal potential forces is equal to the decrease of the internal potential energy of the system, i.e. $\delta A_{int}^{pot} = -dU$. The foregoing expression then takes the form

$$dT + dU = \delta A_{ext} + \delta A_{int}^{dis}. \tag{3.46}$$

Let us introduce the concept of the *total mechanical energy of a system*, or, briefly, *mechanical energy*, as the sum of the kinetic and potential energies of the system:

$$\boxed{E = T + U.} \tag{3.47}$$

Obviously, the energy E depends on the velocities of the particles of the system, the type of interaction between them and the configuration of the system. Besides, the energy E, just as the potential energy U, is defined with an accuracy up to an arbitrary constant, and is a *non-additive* quantity, i.e. the energy E of a system is not equal in the general case to the sum of energies of its individual parts. In accordance with Eq. (3.37)

$$E = \sum E_n + U_{ia}, \tag{3.48}$$

where E_n is the mechanical energy of the nth part of the system and U_{ia} is the potential energy of interactions of its individual parts.

Let us return to Eq. (3.46). We shall rewrite it with an allowance made for Eq. (3.47) in the following form:

$$dE = \delta A_{ext} + \delta A_{int}^{dis}. \tag{3.49}$$

The last expression is valid for an infinitesimal variation of the configuration of the system. In the case of a finite change

$$\boxed{E_2 - E_1 = A_{ext} + A_{int}^{dis},} \tag{3.50}$$

i.e. *the increment of the mechanical energy of a system is equal to the algebraic sum of the works performed by all external forces and all internal dissipative forces.*

Eq. (3.49) may be presented in another form, having divided both its sides by the corresponding time interval dt. Then

$$\boxed{dE/dt = N_{ext} + N_{int}^{dis},} \tag{3.51}$$

i.e. *the time derivative of the mechanical energy of a system is equal to the algebraic sum of powers developed by all external forces and all internal dissipative forces.*

Eqs. (3.49)-(3.51) are valid both in inertial and in non-inertial reference frames. It should be borne in mind that in a non-inertial reference frame one has also to take into account the work (power) of inertial forces acting as external forces, i.e. A_{ext} has to be treated as an algebraic sum of the works performed by the external interaction forces A_{ext}^{ia} and the work performed by the inertial forces A_{in}. To emphasize this fact, we rewrite Eq. (3.50) in the form

$$E_2 - E_1 = A_{in} + A_{ext}^{ia} + A_{int}^{dis}. \tag{3.52}$$

Thus, we have arrived at a significant conclusion: the mechanical energy of a system may vary both due to external forces and due to internal dissipative forces (or, more precisely, due to the algebraic sum of the *works* performed by all these forces). From this another important conclusion follows directly, **the law of conservation of mechanical energy**:

in an inertial reference frame the mechanical energy of a closed system of particles, in which there are no dissipative forces, remains constant in the process of motion, i.e.

$$\boxed{E = T + U = \text{const.}} \tag{3.53}$$

Such a system is called *conservative**. Note that during the motion of a closed conservative system it is the total mechanical energy that remains constant while the kinetic and potential energies vary in the general case. These variations, however, always happen so that the increase of one of them is exactly equal to the decrease in the other, i.e. $\Delta T = -\Delta U$. It is obvious that this is valid only in inertial reference frames.

Furthermore, it follows from Eq. (3.50) that when a closed system is non-conservative, i.e. when there are dissipative forces in it, the mechanical energy of such a system decreases, in accordance with Eq. (3.45):

$$E_2 - E_1 = A_{int}^{dis} < 0. \tag{3.54}$$

It may be stated that the decrease of the mechanical energy is caused by the work performed against the dissipative forces acting in the system. Yet this explanation is formal since it does not reveal the physical nature of dissipative forces.

A more profound examination of the problem has led to the fundamental conclusion about the existence of the **universal law of energy conservation** in nature:

energy is never generated or eliminated, it may only pass from one form into another or be exchanged between individual parts of matter. The concept of energy had to be expanded by introducing some new forms (in addition to mechanical): energy of an electromagnetic field, chemical and nuclear energies, etc.

The universal law of energy conservation thus encompasses even those phenomena for which Newton's laws are not valid. Therefore, it cannot be derived from these laws, but should be treated as an independent law, one of the most extensive generalizations of experimental data.

Returning to Eq. (3.54), we can state the following: every decrease in mechanical energy of a closed system always implies the appearance of the equivalent amount of energy of different kinds which are not associated with *visible* motion. In this respect Eqs. (3.49)-(3.51) can be considered as a more general formulation of the energy conservation

* The solar system can be regarded as a closed conservative system with an adequate degree of accuracy.

law in which the *cause* of the mechanical energy variation for a non-closed system is indicated.

Specifically, the mechanical energy of a non-closed system may remain constant, but this happens only in those cases when, in accordance with Eq. (3.50), the decrease in the energy due to the work performed to overcome internal dissipative forces is equalized by the inflow of energy at the expense of the work performed by external forces.

Finally, we should point out that most problems usually require the energy conservation law to be applied together with the momentum conservation law, or the angular momentum conservation law; sometimes all three laws are used simultaneously. The way this is done will be illustrated in the next two chapters.

Problems to Chapter 3

● **3.1.** A stone of mass m is thrown from the Earth's surface at the angle α to the horizontal with the initial velocity $\mathbf{v}_0$. Ignoring the air drag, find the power developed by the gravity force t seconds after the beginning of the motion, and also the work performed by that force during the first t seconds of the motion.

Solution. The velocity of the stone t seconds after the beginning of the motion is $\mathbf{v} = \mathbf{v}_0 + \mathbf{g}t$. The power developed by the gravity force $m\mathbf{g}$ at that moment is

$$N = m\mathbf{g}\mathbf{v} = m\,(\mathbf{g}\mathbf{v}_0 + g^2 t).$$

In our case $\mathbf{g}\mathbf{v}_0 = gv_0 \cos(\pi/2 + \alpha) = -gv_0 \sin\alpha$, therefore,

$$N = mg\,(gt - v_0 \sin\alpha).$$

It is seen from this that if $t < t_0 = v_0 \sin\alpha/g$, then $N < 0$ while if $t > t_0$, then $N > 0$.

The work performed by the gravity force during the first t seconds is

$$A = \int_0^t N\,dt = mg\,(gt^2/2 - v_0 \sin\alpha \cdot t).$$

The $N(t)$ and $A(t)$ plots are shown in Fig. 47.

● **3.2.** There are two stationary fields of force:

(1) $\mathbf{F} = ay\mathbf{i}$;

(2) $\mathbf{F} = ax\mathbf{i} + by\mathbf{j}$,

where $\mathbf{i}$, $\mathbf{j}$ are the unit vectors of the x and y axes, and a and b are constants. Are these fields of force potential?

Solution. Let us find the work performed by the force of each field over the path from a certain point *1* (x_1, y_1) to a certain point *2* (x_2, y_2):

$$(1) \quad \delta A = \mathbf{F}\, d\mathbf{r} = ay\mathbf{i}\, d\mathbf{r} = ay\, dx; \qquad A = a\int_{x_1}^{x_2} y\, dx;$$

$$(2) \quad \delta A = \mathbf{F}\, d\mathbf{r} = (ax\mathbf{i} + by\mathbf{j})\, d\mathbf{r} = ax\, dx + by\, dy;$$

$$A = a\int_{x_1}^{x_2} x\, dx + b\int_{y_1}^{y_2} y\, dy.$$

In the first case the integral depends on the type of the $y(x)$ function, that is, on the shape of the path. Consequently, the first field of force is not potential. In the second case both integrals do not

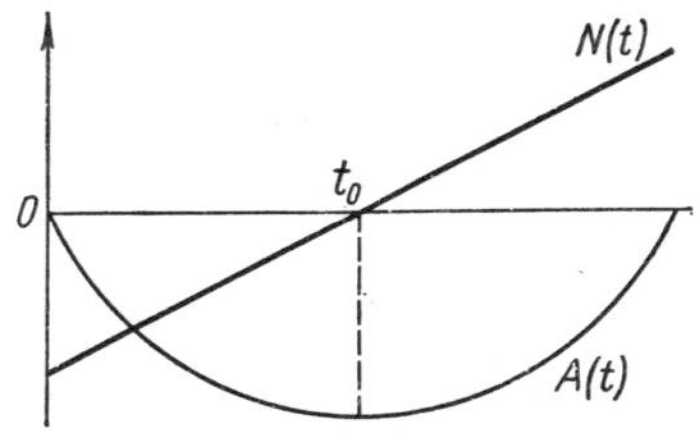

Fig. 47

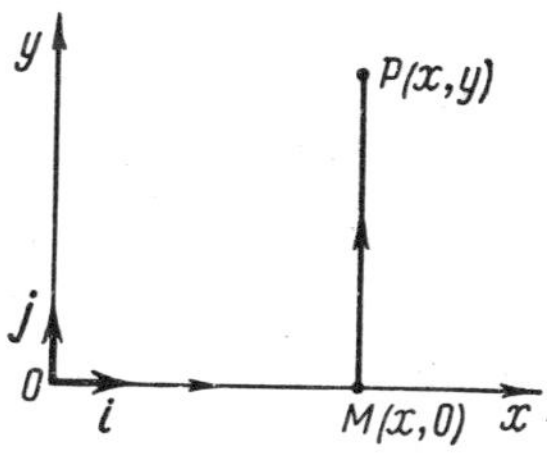

Fig. 48

depend on the shape of the path: they are defined only by the coordinates of the initial and final points of the path; therefore, the second field of force is potential.

● **3.3.** In a certain potential field a particle experiences the force

$$\mathbf{F} = a\,(y\mathbf{i} + x\mathbf{j}),$$

where a is a constant, and $\mathbf{i}$ and $\mathbf{j}$ are the unit vectors of the x and y axes respectively. Find the potential energy $U(x, y)$ of the particle in this field.

Solution. Let us calculate the work performed by the force **F** over the distance from the point O (Fig. 48) to an arbitrary point $P(x, y)$. Taking advantage of this work being independent of the shape of the path, we choose one passing through the points OMP and consisting of two rectilinear sections. Then

$$A_{OP} = \int_0^M \mathbf{F}\, d\mathbf{r} + \int_M^P \mathbf{F}\, d\mathbf{r}.$$

The first integral is equal to zero since at all points of the OM section $y \equiv 0$ and $\mathbf{F} \perp d\mathbf{r}$. Along the section MP $x =$ const, $\mathbf{F}\, d\mathbf{r} = \mathbf{F}\mathbf{j}\, dy =$

$= F_y\, dy = ax\, dy$ and therefore,

$$A_{OP} = 0 + ax \int_M^P dy = axy.$$

In accordance with Eq. (3.10) this work must be equal to the decrease in the potential energy, i.e. $A_{OP} = U_O - U_P$. Assuming $U_O = 0$, we obtain $U_P = -A_{OP}$, or

$$U(x, y) = -axy.$$

Another way of finding $U(x, y)$ is to resort to the total differential of that function:

$$dU = (\partial U/\partial x)\, dx + (\partial U/\partial y)\, dy.$$

Taking into account that $\partial U/\partial x = -F_x = -ay$ and $\partial U/\partial y = -F_y = -ax$, we get

$$dU = -a(y\, dx + x\, dy) = d(-axy).$$

Whence $U(x, y) = -axy$.

● **3.4.** A ball of mass m is suspended on a weightless elastic thread whose stiffness factor is equal to $\varkappa$. Then the ball is lifted so that the thread is not stretched, and let fall with the zero initial velocity. Find the maximum stretch x_m of the thread in the process of the ball's motion.

Solution. Let us consider several solution methods based on the energy conservation law.

1. According to Eq. (3.28) the increment of the kinetic energy of the ball is equal to the algebraic sum of the works performed by *all* forces acting on it. In our case those are the gravity force mg and the elastic force $F_{el} = \varkappa x$ of the thread (Fig. 49*a*). In the initial and final positions of the ball its kinetic energy is equal to zero for when the maximum stretch occurs the ball stops moving. Therefore, in accordance with Eq. (3.28) the sum of the works $A_{gr} + A_{el} = 0$, or

$$mgx_m + \int_0^{x_m} (-\varkappa x)\, dx = mgx_m - \varkappa x_m^2/2 = 0.$$

Whence $x_m = 2mg/\varkappa$.

2. The ball may be considered in the Earth's gravity field. This approach requires the total mechanical energy of the ball in the Earth's gravity field to be analysed. In accordance with Eq. (3.32) the increment of this energy is equal to the work performed by the external forces. In this case it is the elastic force that should be treated as an external force while the increment of the total mechanical energy of the ball is equal to that of only its potential energy in the Earth's

gravity field. Therefore,

$$0 - mgx_m = \int_0^{x_m} (-\varkappa x)\, dx = -\varkappa x_m^2/2.$$

Whence we get the same result for x_m.

Note that we could proceed in the opposite way, examining the ball in the field of the elastic force and treating the gravity force as an external one. It pays to make sure that the result obtained in that approach is the same.

3. And finally, we can consider the ball in the field generated by the joint action of both the gravity force and the elastic force. Then external forces are absent, and the total mechanical energy of the

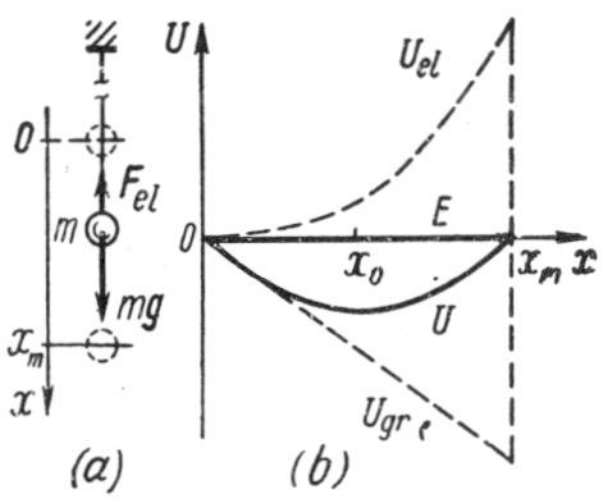

Fig. 49

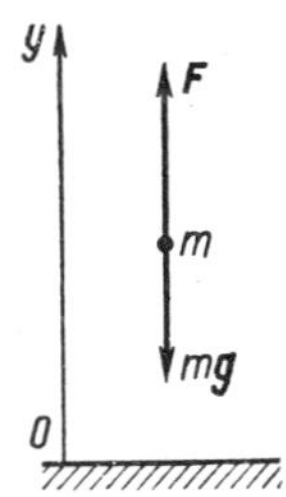

Fig. 50

ball in such a field remains constant in the process of motion, i.e. $\Delta E = \Delta T + \Delta U = 0$. During the transition of the ball from the initial to the final (lower) position $\Delta T = 0$, and therefore $\Delta U = \Delta U_{gr} + \Delta U_{el} = 0$, or

$$-mgx_m + \varkappa x_m^2/2 = 0.$$

Again we get the same result.

Fig. 49*b* shows the plots $U_{gr}(x)$ and $U_{el}(x)$, whose origins are assumed to be located at the point $x = 0$ (which is not mandatory, of course). The same figure illustrates the plot of the total potential energy $U(x) = U_{gr} + U_{el}$. For a given choice of the potential energy reference value the total mechanical energy of the ball $E = 0$.

● **3.5.** A body of mass m ascends from the Earth's surface with zero initial velocity due to the action of the two forces (Fig. 50): the force $\mathbf{F}$ varying with the height y as $\mathbf{F} = -2m\mathbf{g}(1 - ay)$, where a is a positive constant, and the gravity $m\mathbf{g}$. Find the work performed by the force $\mathbf{F}$ over the first half of the ascent and the corresponding increment of the potential energy of the body in the Earth's gravity field, which is assumed to be uniform.

Solution. First, let us find the total height of ascent. At the beginning and the end of the path the velocity of the body is equal to zero, and therefore the increment of the kinetic energy of the body is also equal to zero. On the other hand, in accordance with Eq. (3.28) ΔT is equal to the algebraic sum of the works A performed by all the forces, i.e. by the force **F** and gravity, over this path. However, since

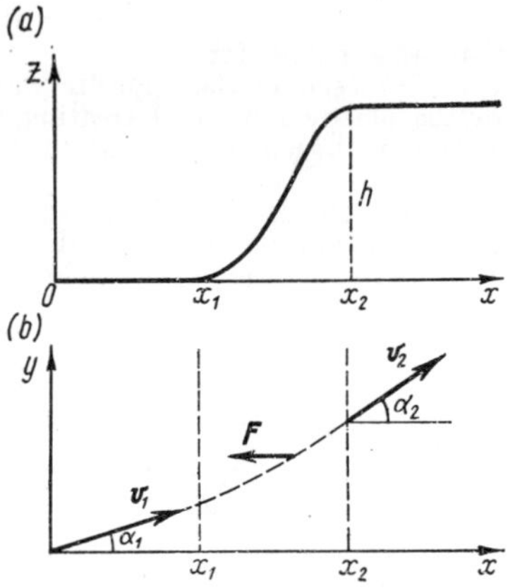

Fig. 51

$\Delta T = 0$, then $A = 0$. Taking into account that the upward direction is assumed to coincide with the positive direction of the y axis, we can write

$$A = \int_0^h (\mathbf{F} + m\mathbf{g})\, d\mathbf{r} = \int_0^h (F_y - mg)\, dy =$$

$$= mg \int_0^h (1 - 2ay)\, dy = mgh\,(1 - ah) = 0.$$

Whence $h = 1/a$.

The work performed by the force **F** over the first half of the ascent is

$$A_F = \int_0^{h/2} F_y\, dy = 2mg \int_0^{h/2} (1 - ay)\, dy = 3mg/4a.$$

The corresponding increment of the potential energy is

$$\Delta U = mgh/2 = mg/2a.$$

● **3.6.** A disc slides without friction up a hill of height h whose profile depends only on the x coordinate (Fig. 51a). At the bottom the disc has the velocity v_1 whose direction forms the angle α_1 with the x axis (see Fig. 51*b*, top view). Find the motion direction of the disc after it reaches the top, i.e. find the angle α_2.

Solution. First of all we shall note that this problem cannot be solved by the use of the fundamental equation of dynamics since the force F acting on the disc in the region $x_1 < x < x_2$ is not specified. All we know about this force is that it is perpendicular to the y axis.

We shall make use of the energy conservation law: $mv_1^2 = mv_2^2 + 2mgh$, whence

$$v_2^2 = v_1^2 - 2gh. \quad (1)$$

This expression can be rewritten as follows:

$$v_{2x}^2 + v_{2y}^2 = v_{1x}^2 + v_{1y}^2 - 2gh.$$

Since the force of the field is perpendicular to the y axis, it does not affect the v_y projection of the velocity; hence, $v_{2y} = v_{1y}$. Therefore, the previous expression may be reduced to $v_{2x}^2 = v_{1x}^2 - 2gh$, or

$$v_2 \cos \alpha_2 = \sqrt{v_1^2 \cos^2 \alpha_1 - 2gh}, \quad (2)$$

where v_2 is defined from Eq. (1). As a result

$$\cos \alpha_2 = \sqrt{(v_1^2 \cos^2 \alpha_1 - 2gh)/(v_1^2 - 2gh)}.$$

Note that this expression holds if the radicand in Eq. (2) is not negative, i.e. when $v_1 \cos \alpha_1 > \sqrt{2gh}$. Otherwise the disc cannot overcome the hill, that is, it is "reflected" from the potential barrier.

● **3.7.** A plane spiral made of stiff smooth wire is rotated with a constant angular velocity ω in a horizontal plane about the fixed vertical axis O (Fig. 52). A small sleeve M slides along that spiral without friction. Find its velocity v' relative to the spiral as a function of the distance ρ from the rotation axis O if the initial velocity of the sleeve is equal to v_0'.

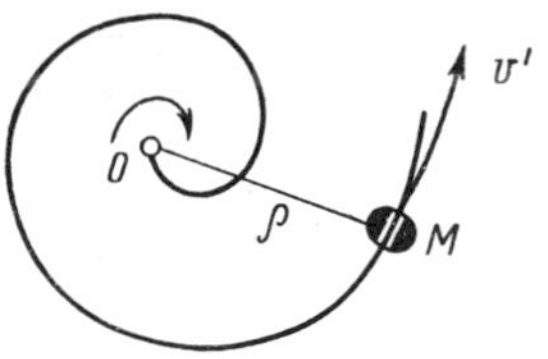

Fig. 52

Solution. It is advisable to solve this problem in a reference frame fixed to the spiral. We know that the increment of the kinetic energy of the sleeve must be equal to the algebraic sum of the works performed by all the forces acting on it. It can easily be reasoned that of all forces work is performed only by the centrifugal force of inertia. The remaining forces, gravity, the force of reaction of the spiral, the Coriolis force, are perpendicular to the **v**′ velocity of the sleeve and do not perform any work.

In accordance with Eq. (3.28),

$$m(v'^2 - v_0'^2)/2 = \int m\omega^2 \boldsymbol{\rho}\, d\mathbf{r},$$

where m is the mass of the sleeve and $d\mathbf{r}$ is its elementary displacement relative to the spiral. Since $\boldsymbol{\rho}\, d\mathbf{r} = \rho\, (d\mathbf{r})_\rho = \rho\, d\rho$, the integral proves to be equal to $m\omega^2\rho^2/2$. Hence, the sought velocity

$$v' = \sqrt{v_0'^2 + \omega^2\rho^2}.$$

● 3.8. Find the strength and the potential of the gravitational field generated by a uniform sphere of mass M and radius R as a function of the distance r from its centre.

Solution. First of all we shall demonstrate that the potential produced by a thin uniform spherical layer of substance outside the layer is such as if the whole mass of the layer were concentrated at its centre, while the potential within the layer is the same at all its points.

Suppose the thin spherical layer has the mass m and radius a. Let us start with calculating the potential $d\varphi$ at the point P $(r > a)$ forming the elementary band dS of the given layer (Fig. 53*a*). If the

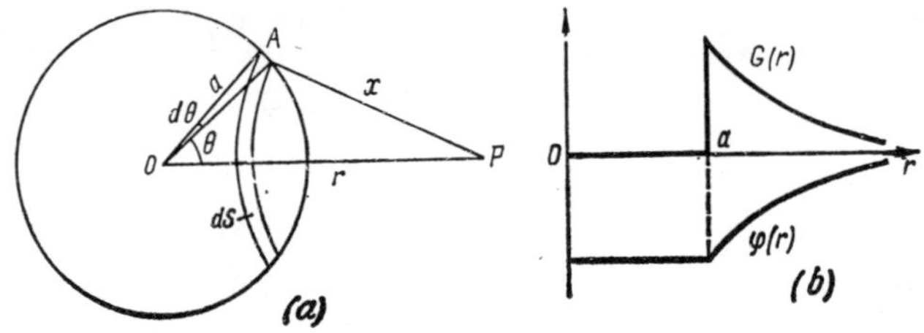

Fig. 53

mass of this band is equal to dm and all its points are located at the distance x from the point P, then $d\varphi = -\gamma\, dm/x$. Allowing for $dm = (m/2) \sin\theta\, d\theta$, we obtain

$$d\varphi = -(\gamma m/2x) \sin\theta\, d\theta. \tag{1}$$

Next, from the cosine theorem (for the triangle OAP) it follows that $x^2 = a^2 + r^2 - 2ar\cos\theta$. Taking the differential of this expression, we get

$$x\, dx = ar \sin\theta\, d\theta. \tag{2}$$

Using Eq. (2) we reduce Eq. (1) to the form $d\varphi = -(\gamma m/2ar)\, dx$; then we integrate this equation over the whole layer. Then

$$\varphi_{outside} = -(\gamma m/2ar) \int_{r-a}^{r+a} dx = -\gamma m/r. \tag{3}$$

Thus, the potential at the point P outside the thin uniform spherical layer is indeed such as if the whole mass of that layer were concentrated at its centre.

When the point P is located inside the layer $(r < a)$, the foregoing calculations remain valid till the integration. In this case the integration with respect to x is to be carried out from $a - r$ to $a + r$. As a result,

$$\varphi_{inside} = -\gamma m/a, \tag{4}$$

that is, the potential inside the layer does not depend on the position of the point P, and consequently it will be the same at all points inside the layer.

In accordance with Eq. (3.24) the field strength at the point P is equal to

$$G_r = -\frac{\partial \varphi}{\partial r} = \begin{cases} -\gamma m/r^2 & \text{outside the layer,} \\ 0 & \text{inside the layer.} \end{cases}$$

The plots $\varphi(r)$ and $G(r)$ for a thin spherical layer are illustrated in Fig. 53*b*.

Now let us generalize the results obtained to a uniform sphere of mass M and radius R. If the point P lies outside the sphere ($r > R$), then from Eq. (3) it immediately follows that

$$\varphi_{outside} = -\gamma M/r. \tag{5}$$

But if the point P lies inside the sphere ($r < R$), the potential φ at that point may be represented as a sum:

$$\varphi = \varphi_1 + \varphi_2,$$

where φ_1 is the potential of a sphere having the radius r, and φ_2 is the potential of the layer of radii r and R. In accordance with Eq. (5),

$$\varphi_1 = -\gamma \frac{M (r/R)^3}{r} = -\gamma \frac{M}{R^3} r^2.$$

The potential φ_2 produced by the layer is the same at all points inside it. The potential φ_2 is easiest to calculate for the point positioned at the layer's centre:

$$\varphi_2 = -\gamma \int_r^R \frac{dM}{r} = -\frac{3}{2} \frac{\gamma M}{R^3} (R^2 - r^2),$$

where $dM = (3M/R^3)\, r^2\, dr$ is the mass of a thin layer between the radii r and $r + dr$. Thus,

$$\varphi_{inside} = \varphi_1 + \varphi_2 = -(\gamma M/2R)(3 - r^2/R^2). \tag{6}$$

The field strength at the point P follows from Eqs. (5) and (6):

$$G_r = -\frac{\partial \varphi}{\partial r} = \begin{cases} -\gamma M/r^2 & \text{with } r > R, \\ -\gamma M r/R^3 & \text{with } r < R. \end{cases}$$

The plots $\varphi(r)$ and $G(r)$ for a uniform sphere of radius R are shown in Fig. 54.

● 3.9. Demonstrate that the kinetic energy T_2 which a body requires to escape the Earth's gravitational pull is twice as high as the energy T_1 required to launch that body into a circular orbit around the Earth (close to its surface). The air drag and the Earth's rotation are to be ignored.

Solution. Let us find the velocity v_1 of a body travelling along a circular orbit. In accordance with the fundamental equation of dynamics

$$mv_1^2/R = mg,$$

where m is the mass of the body, and R is the orbit radius equal approximately to the Earth's radius. Hence,

$$v_1 = \sqrt{gR} = 7.9 \text{ km/s}.$$

This is the so-called *first cosmic velocity*.

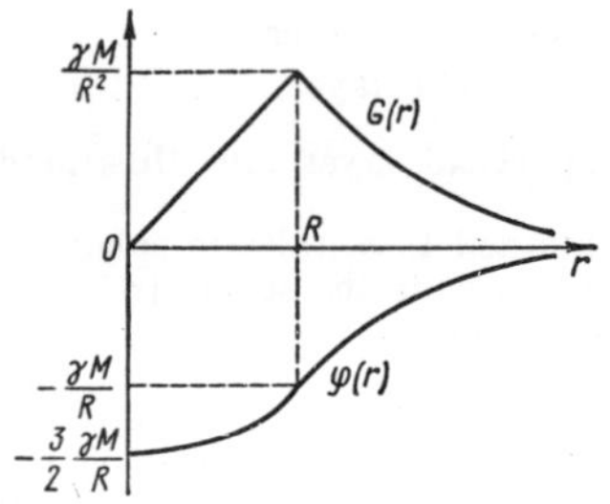

Fig. 54

To overcome the Earth's gravitational pull, the body has to reach the *second cosmic velocity* v_2. Its magnitude can be found from the energy conservation law: the kinetic energy of the body close to the Earth's surface must be equal to the height of the potential barrier that the body must overcome. The height of the barrier is equal to the increment of the potential energy of the body between the points $r = R$ and $r = \infty$. Thus,

$$mv_2^2/2 = \gamma mM/R,$$

where M is the Earth's mass. Hence,

$$v_2 = \sqrt{2\gamma M/R} = \sqrt{2gR} = 11 \text{ km/s}.$$

Consequently, $v_2 = \sqrt{2}\, v_1$ and $T_2 = 2T_1$.

● **3.10.** Three identical charged particles, each possessing the mass m and charge $+q$, are placed at the corners of an equilateral triangle with the side r_0. Then the particles are simultaneously set free and start flying apart symmetrically due to Coulomb's repulsion forces. Find:

(1) the velocity of each particle as a function of the distance r between them;

(2) the work A_1 performed by Coulomb's forces acting on each particle until the particles fly from one another to a very large distance.

Solution. 1. Since the given system is closed, the increment of its kinetic energy is equal to the decrease in the potential energy, i.e.

$$3mv^2/2 = 2kq^2/r_0 - 3kq^2/r.$$

Hence,

$$v = \sqrt{2kq^2(r - r_0)/mrr_0}.$$

It is seen that as $r \to \infty$ the velocity of each particle approaches the limiting value $v_{max} = \sqrt{2kq^2/mr_0}$.

2. The work performed by the interaction forces during the variation of the system's configuration is equal to the decrease in the potential energy of the system:

$$A = U_1 - U_2 = 3kq^2/r_0,$$

where allowance is made for the fact that at the final position $U_2 = 0$. Hence, the sought work is

$$A_1 = A/3 = kq^2/r_0. \qquad (1)$$

Note. Here attention should be drawn to a frequent mistake made when solving this kind of problem. In particular, one may often hear the following argument: at the initial position the potential energy of each particle is equal to $2kq^2/r_0$ while at the final position to zero. Hence, the sought work $A_1 = 2kq^2/r_0$. This result is twice as great as the one of Eq. (1). Why is this so?

The mistake is made because the field in which each particle travels is non-stationary and, consequently, non-potential (since the two other particles also displace relative to each other), and therefore the work in such a field cannot be represented as a decrease in the potential energy of the particle.

● **3.11.** Making use of the results obtained in solving Problem 3.8, find the internal potential energy of gravitational interaction of masses distributed uniformly:

(1) over the surface of a sphere;

(2) throughout the volume of a sphere.

The mass of the sphere is equal to M and its radius to R.

Solution. 1. Since the potential at each point of a spherical surface is $\varphi = -\gamma M/R$, we obtain, in accordance with Eq. (3.40):

$$U = (\varphi/2) \int dm = -\gamma M^2/2R.$$

2. In this case the potential inside the sphere depends only on r (see Problem 3.8):

$$\varphi = -(3\gamma M/2R)(1 - r^2/3R^2).$$

Substituting this expression into Eq. (3.40) and integrating, we get

$$U = \frac{1}{2} \int_{r=0}^{R} \varphi \, dm = -\frac{3}{5} \frac{\gamma M^2}{R},$$

where dm is the mass of an elementary spherical layer confined between the radii r and $r + dr$; $dm = (3M/R^3) r^2 \, dr$.

CHAPTER 4

THE LAW OF CONSERVATION OF MOMENTUM

§ 4.1. Momentum. The Law of Its Conservation

The momentum* of a particle. Practical knowledge and analysis of mechanical phenomena indicate that apart from the kinetic energy $T = mv^2/2$ one needs to introduce one more quantity, *momentum* ($\mathbf{p} = m\mathbf{v}$), in order to describe the mechanical motion of bodies. These quantities provide the basic measures of mechanical motion of bodies, the former being scalar and the latter vectorial. Both of them play a most significant part in the construction of mechanics.

Let us proceed to a more detailed analysis of momentum. First of all we shall write the fundamental equation of Newtonian dynamics (2.6) in another form, by the use of momentum:

$$\boxed{d\mathbf{p}/dt = \mathbf{F},} \tag{4.1}$$

i.e. *the time derivative of the momentum of a mass point is equal to the force acting on that point.* Specifically, if $\mathbf{F} \equiv 0$, then $\mathbf{p} = \text{const}$.

Note that in a non-inertial reference frame the force $\mathbf{F}$ comprises not only the forces of interaction between a given particle and other bodies, but also inertial forces.

Eq. (4.1) allows the increment of the momentum of a particle to be found for any time interval provided the time dependence of the force $\mathbf{F}$ is known. In fact, it follows from Eq. (4.1) that the elementary momentum increment that the particle acquires during the time interval dt is equal to $d\mathbf{p} = \mathbf{F}\, dt$. Integrating this expression with respect to time, we find the momentum increment of a particle during the finite time interval t:

$$\mathbf{p}_2 - \mathbf{p}_1 = \int_0^t \mathbf{F}\, dt. \tag{4.2}$$

---

* It is sometimes called the *quantity of motion.*

When the force $\mathbf{F}$ = const, the vector $\mathbf{F}$ can be removed from the integrand and then $\mathbf{p}_2 - \mathbf{p}_1 = \mathbf{F}t$. The quantity on the right-hand side of this equation is referred to as the *power impulse*. Thus, the momentum increment acquired by a particle during any time interval is equal to the power impulse over the same time interval.

Example. A particle which at the initial moment $t = 0$ possesses the momentum $\mathbf{p}_0$ is subjected to the force $\mathbf{F} = \mathbf{a}t\,(1 - t/\tau)$ during the time interval τ, $\mathbf{a}$ being a constant vector. Find the momentum $\mathbf{p}$ of the particle at the moment when the action of the force comes to an end.

In accordance with Eq. (4.2) we get $\mathbf{p} = \mathbf{p}_0 + \int_0^\tau \mathbf{F}\,dt = \mathbf{p}_0 + \mathbf{a}\tau^2/6$ (Fig. 55).

The momentum of a system. Let us consider an arbitrary system of particles and introduce the concept of the *momentum of a system* as a vector sum of the momenta of its constituent particles:

$$\mathbf{p} = \sum \mathbf{p}_i, \qquad (4.3)$$

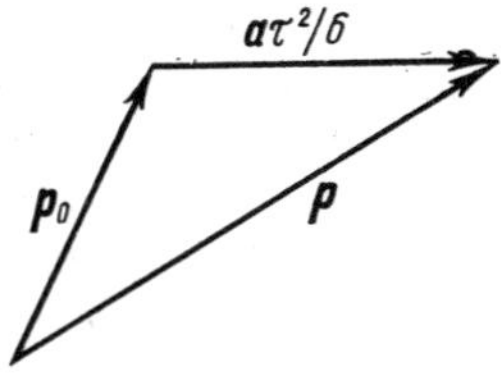

Fig. 55

where $\mathbf{p}_i$ is the momentum of the ith particle. Note that the momentum of a system is an additive quantity, that is, the momentum of a system is equal to the sum of the momenta of its individual parts irrespective of whether or not they interact.

Let us find the physical quantity which defines the system's momentum change. For this purpose we shall differentiate Eq. (4.3) with respect to time:

$$d\mathbf{p}/dt = \sum d\mathbf{p}_i/dt.$$

In accordance with Eq. (4.1)

$$d\mathbf{p}_i/dt = \sum_k \mathbf{F}_{ik} + \mathbf{F}_i,$$

where $\mathbf{F}_{ik}$ are the forces which the other particles of the system exert on the ith particle, i.e. internal forces; $\mathbf{F}_i$ is the force which other bodies outside the system under consideration exert on the same particle (external forces). Substituting the last expression into the previous one, we get

$$d\mathbf{p}/dt = \sum_i \sum_k \mathbf{F}_{ik} + \sum_i \mathbf{F}_i.$$

The double summation symbol on the right-hand side denotes the sum of all internal forces. In accordance with Newton's third law the interaction forces between the system's particles are pairwise identical in magnitude and opposite in direction. Consequently, the resultant force of each interacting pair is equal to zero, and therefore the vector sum of all internal forces is also equal to zero. As a result, the last equation takes the following form:

$$\boxed{d\mathbf{p}/dt = \mathbf{F},} \tag{4.4}$$

where $\mathbf{F}$ is the resultant of all *external* forces, $\mathbf{F} = \sum \mathbf{F}_i$.

Eq. (4.4) implies that the *time derivative of the momentum of a system is equal to the vector sum of all external forces* acting on the particles of the system.

As in the case of a single particle, it follows from Eq. (4.4) that the increment of momentum which the system acquires during the finite time interval t is equal to

$$\boxed{\mathbf{p}_2 - \mathbf{p}_1 = \int_0^t \mathbf{F}\,dt,} \tag{4.5}$$

i.e. the increment of momentum of the system is equal to the momentum of the resultant of all external forces over the corresponding time interval. Here $\mathbf{F}$ is the resultant of all the *external* forces.

Eqs. (4.4) and (4.5) hold true both in inertial and in non-inertial reference frames. It should be borne in mind, however, that in non-inertial reference frames one needs to take into account the *inertial forces*, which act as external forces, i.e. in these equations $\mathbf{F}$ should be regarded as the sum $\mathbf{F}_{ia} + \mathbf{F}_{in}$, where $\mathbf{F}_{ia}$ is the resultant of all external interaction forces, and $\mathbf{F}_{in}$ is the resultant of all inertial forces.

The law of momentum conservation. We have drawn an important conclusion: in accordance with Eq. (4.4) *the momentum of a system may vary only due to external forces.* Internal forces cannot change the momentum of a system. Hence, another important conclusion immediately follows from this, the **law of momentum conservation:** *in an inertial reference frame the momentum of a closed system of particles remains constant*, i.e. does not change in the course of time:

$$\boxed{\mathbf{p} = \sum \mathbf{p}_i\ (t) = \text{const.}} \tag{4.6}$$

Here the momenta of individual particles or parts of a closed system may change with time, a fact emphasized in the last expression. These changes, however, always happen so that the momentum increment of one part of the system is equal to the momentum decrease of another part of the system. In other words, the individual parts of a closed system can only interchange momenta. Having detected a momentum increment in a certain system, we can state that this increment originated at the expense of a momentum decrease in surrounding bodies.

In this regard Eqs. (4.4) and (4.5) should be treated as a more general formulation of the momentum conservation law. This formulation indicates that the momentum change of a non-closed system is caused by the action of other bodies (external forces). What was said is of course valid only in reference to inertial reference frames.

The momentum of a non-closed system can remain constant provided the resultant of all external forces is equal to zero. This follows immediately from Eqs. (4.4) and (4.5). In these cases the conservation of momentum is of practical interest, for it permits the system to be studied in a sufficiently simple fashion without going into a detailed analysis of the process.

One more thing. Sometimes in a non-closed system it is not the momentum $\mathbf{p}$ itself that remains constant, but its p_x projection on a certain x direction. This happens when the projection of the resultant $\mathbf{F}$ of the external forces on the x direction is equal to zero, i.e. the vector $\mathbf{F}$ is perpendicular to that direction. In fact, projecting Eq. (4.4), we get

$$dp_x/dt = F_x, \tag{4.7}$$

whence it follows that if $F_x \equiv 0$, then $p_x = \text{const}$. For example, when a system moves in a uniform field of gravity, the projection of its momentum on any horizontal direction remains constant whatever happens inside the system.

Let us consider examples involving constant and varying momenta.

Example 1. A cannon of mass m slides down a smooth inclined plane forming the angle α with the horizontal. At the moment when the velocity of the cannon reaches $\mathbf{v}$, it fires a shell in a horizontal direction with the result that the cannon stops and the shell "carries away" the momentum $\mathbf{p}$. Suppose that the firing duration is equal to t. What is the reaction force $\mathbf{R}$ of the inclined plane averaged over the time t?

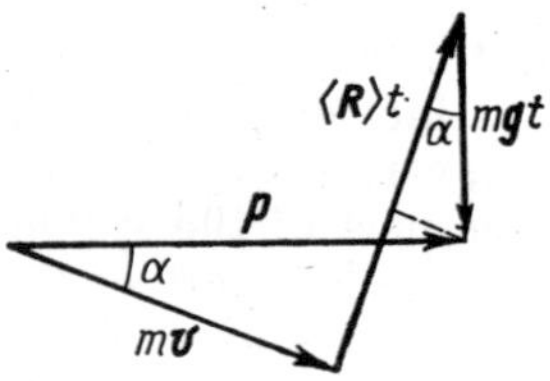

Fig. 56

Here the system "cannon-shell" is non-closed. During the time interval t this system acquires a momentum increment equal to $\mathbf{p} - m\mathbf{v}$. The change of the system's momentum is caused by two external forces: the reaction force $\mathbf{R}$ (which is perpendicular to the inclined plane) and gravity $m\mathbf{g}$. Therefore, we can write:

$$\mathbf{p} - m\mathbf{v} = \langle \mathbf{R} \rangle t + m\mathbf{g}t,$$

where $\langle \mathbf{R} \rangle$ is the vector $\mathbf{R}$ averaged over the time t. It is helpful to depict this relationship graphically (Fig. 56). It can be immediately seen from the figure that the sought value $\langle R \rangle$ is defined by the formula $\langle R \rangle t = p \sin \alpha + mgt \cos \alpha$.

Example 2. A man of mass m_1 is located on a narrow raft of mass m_2 afloat on the surface of a lake. The man travels through the distance $\Delta\mathbf{r}'$ with respect to the raft and then stops. The resistance of the water is negligible. We shall find the corresponding displacement $\Delta\mathbf{r}_2$ of the raft relative to the shore.

In this case the resultant of all external forces acting on the "man-raft" system is equal to zero, and, therefore, the momentum of that system does not change, remaining equal to zero in the process of motion:

$$m_1\mathbf{v}_1 + m_2\mathbf{v}_2 = 0,$$

where $\mathbf{v}_1$ and $\mathbf{v}_2$ are the velocities of the man and the raft with respect to the shore. But the velocity of the man relative to the shore may be represented in the form $\mathbf{v}_1 = \mathbf{v}_2 + \mathbf{v}'$, where $\mathbf{v}'$ is the velocity of the man relative to the raft. Eliminating $\mathbf{v}_1$ from these two equations, we obtain

$$\mathbf{v}_2 = -\frac{m_1}{m_1+m_2}\mathbf{v}'.$$

Multiplying both sides by dt, we find the relationship between the elementary displacements of the raft $d\mathbf{r}_2$ and the man $d\mathbf{r}'$ relative to the raft. Obviously, the same relationship also holds in the case of finite displacements:

$$\Delta\mathbf{r}_2 = -\frac{m_1}{m_1+m_2}\Delta\mathbf{r}'.$$

It is seen from this that the displacement $\Delta\mathbf{r}_2$ of the raft does not depend on the character of the man's motion, i.e. on the law $\mathbf{v}'(t)$.

We emphasize once again that the momentum conservation law holds only in inertial frames. This statement, however, does not rule out the momentum of a system remaining constant in non-inertial reference frames as well. This happens when in Eq. (4.4), which is also valid in non-inertial reference frames, the external force $\mathbf{F}$ (including inertial forces) is equal to zero. Clearly this situation occurs only under special conditions. Such special cases are fairly rare.

Let us now demonstrate that if the momentum of a system remains constant in one inertial reference frame K, it also does so in any other inertial frame K'. Suppose in the K frame

$$\sum m_i\mathbf{v}_i = \text{const}.$$

If the K' frame moves relative to the K frame with the velocity $\mathbf{V}$, the velocity of the ith particle in the K frame may be written as $\mathbf{v}_i = \mathbf{v}'_i + \mathbf{V}$, where $\mathbf{v}'_i$ is the velocity of that particle in the K' frame. Then the expression for the momentum of the system may be transformed as follows: $\sum m_i\mathbf{v}'_i + \sum m_i\mathbf{V} = \text{const}$. The second term here does not depend on time. This implies that the first term, the momentum of the system in the K' reference frame, does not depend on time either, i.e.

$$\sum m_i\mathbf{v}'_i = \text{const}'.$$

The result obtained is in complete agreement with the Galilean relativity principle, according to which the laws of mechanics are identical in all inertial reference frames.

The validity of Newton's laws underlies the reasoning that led us to the momentum conservation law. Specifically, the mass points of a closed system were assumed to interact in pairs and to obey Newton's third law. Now, what happens in systems which do not obey Newton's laws, e.g. in systems involving electromagnetic radiation?

Experience shows convincingly enough that the momentum conservation law is valid for such systems as well. However, in these cases one has to take into account in the general equilibrium of momenta not only the momenta of particles, but also the momentum which, as electrodynamics confirms, the radiation field itself possesses.

Thus, experience shows that the *momentum conservation law*, when appropriately correlated, *constitutes a fundamental law of nature which is valid without exceptions.* But in this broad sense, this law is no longer a consequence of Newton's laws, and should be regarded as an independent general principle inferred from experimental data.

§ 4.2. Centre of Inertia. The C Frame.

The centre of inertia. Any system of particles possesses one remarkable point C, the *centre of inertia*, or the *centre of mass*, displaying a number of interesting and significant properties. Its position relative to the origin O of a given reference frame is described by the radius vector $\mathbf{r}_C$ defined by the following formula:

$$\mathbf{r}_C = \frac{1}{m} \sum m_i \mathbf{r}_i, \tag{4.8}$$

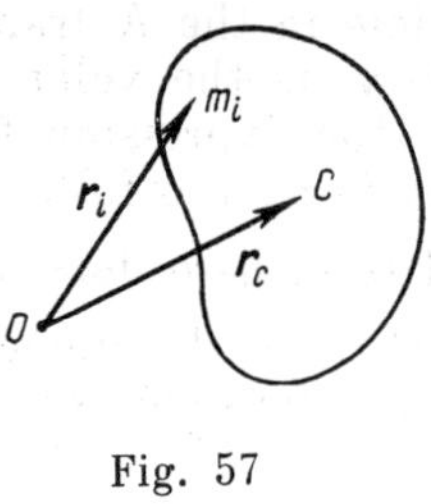

Fig. 57

where m_i and $\mathbf{r}_i$ are the mass and the radius vector of the ith particle, and m is the mass of the whole system (Fig. 57).

It should be pointed out that the centre of inertia of a system coincides with its centre of gravity. However, this statement is valid only when the gravitational field can be assumed uniform within the limits of a given system.

Let us find the velocity of the centre of inertia in a given reference frame. Differentiating Eq. (4.8) with respect to time, we get

$$\mathbf{V}_C = \frac{1}{m} \sum m_i \mathbf{v}_i = \frac{1}{m} \sum \mathbf{p}_i. \tag{4.9}$$

If the velocity of the centre of inertia is equal to zero, the system is said to be at rest as a whole. This provides a natural generalization of the concept of a motionless particle. Accordingly, the velocity $\mathbf{V}_C$ acquires the meaning of the velocity of the system moving as a whole.

With allowance made for Eq. (4.3) we obtain from Eq. (4.9)

$$\mathbf{p} = m\mathbf{V}_C, \tag{4.10}$$

i.e. *the momentum of a system is equal to the product of the mass of the system by the velocity of its centre of inertia.*

The equation of motion for the centre of inertia. The concept of a centre of inertia allows Eq. (4.4) to be rewritten in a more convenient form. To do this, we have to substitute Eq. (4.10) into Eq. (4.4) and take into account that the mass of a system *per se* has a constant value. Then we obtain

$$\boxed{m \frac{d\mathbf{V}_C}{dt} = \mathbf{F},} \tag{4.11}$$

where $\mathbf{F}$ is the resultant of all external forces acting on the system. This is the *equation of motion for the centre of inertia* of a system, one of the most important equations of mechanics. According to this equation, *during the motion of any system of particles its centre of inertia moves as if all the mass of the system were concentrated at that point, and all external forces acting on the system were applied to it.* In this case the acceleration of the centre of inertia is quite independent of the points to which the external forces are applied.

Next, it follows from Eq. (4.11) that if $\mathbf{F} \equiv 0$, then $d\mathbf{V}_C/dt \equiv 0$, and therefore $\mathbf{V}_C = \text{const}$. In particular, this case is realized in a closed system (in an inertial reference frame). Furthermore, if $\mathbf{V}_C = \text{const}$, then in accordance with Eq. (4.10) the momentum of the system $\mathbf{p} = \text{const}$.

Thus, *if the centre of inertia of a system moves uniformly and rectilinearly, the momentum of the system remains con-*

stant in the process of motion. Obviously, the reverse statement is also true.

Eq. (4.11) coincides in its form with the fundamental equation of dynamics of a mass point and is its natural generalization to a system of particles: the acceleration of a system as a whole is proportional to the resultant of all external forces and inversely proportional to the total mass of the system. Recall that in non-inertial reference frames the resultant of all external forces includes both forces of interaction with surrounding bodies and inertial forces.

Let us consider three examples associated with motion of a system's centre of inertia.

Example 1. We shall show how the problem of a man on a raft (see Example 2 on p. 118) can be solved by resorting to the notion of the centre of inertia.

Since the resistance of water is negligibly small, the resultant of all external forces acting on the system "a man and a raft" is equal to zero. This means that the position of the centre of inertia of the given system does not change in the process of motion of the man (and the raft), i.e.

$$m_1\mathbf{r}_1 + m_2\mathbf{r}_2 = \text{const},$$

where $\mathbf{r}_1$ and $\mathbf{r}_2$ are the radius vectors describing the positions of the centres of inertia of the man and the raft relative to a certain point on the shore. From this equality we find the relationship between the increments of the vectors $\mathbf{r}_1$ and $\mathbf{r}_2$:

$$m_1\,\Delta\mathbf{r}_1 + m_2\,\Delta\mathbf{r}_2 = 0.$$

Taking into account that the increments $\Delta\mathbf{r}_1$ and $\Delta\mathbf{r}_2$ represent the displacements of the man and the raft with respect to the shore and that $\Delta\mathbf{r}_1 = \Delta\mathbf{r}_2 + \Delta\mathbf{r}'$, we find the displacement of the raft:

$$\Delta\mathbf{r}_2 = -\frac{m_1}{m_1+m_2}\,\Delta\mathbf{r}'.$$

Example 2. A man jumps down from a tower into water. In the general case his motion is quite complicated. However, if the air drag is negligible, it can be immediately stated that the centre of inertia of the jumper moves along a parabola, just as a mass point experiencing the constant force $m\mathbf{g}$, where m is the man's mass.

Example 3. A closed chain connected by a thread to a rotating shaft revolves around a vertical axis with the uniform angular velocity ω (Fig. 58), the thread forming the angle θ with the vertical. How does the centre of inertia of the chain move?

First of all, it is clear that it does not move in the vertical direction during the uniform rotation of the chain. This means that the vertical component of the tensile strength T of the thread counterbalances gravity (Fig. 58, right). As for the horizontal component of the tensile strength, it is constant in magnitude and permanently directed toward the rotation axis. It follows from this that the centre of inertia of the chain, the point C, travels along the horizontal circle whose radius ρ is easy to find via Eq. (4.11), writing it as

$$m\omega^2\rho = mg \tan \theta,$$

where m is the mass of the chain. In this case the point C is permanently located between the rotation axis and the thread, as shown in Fig. 58.

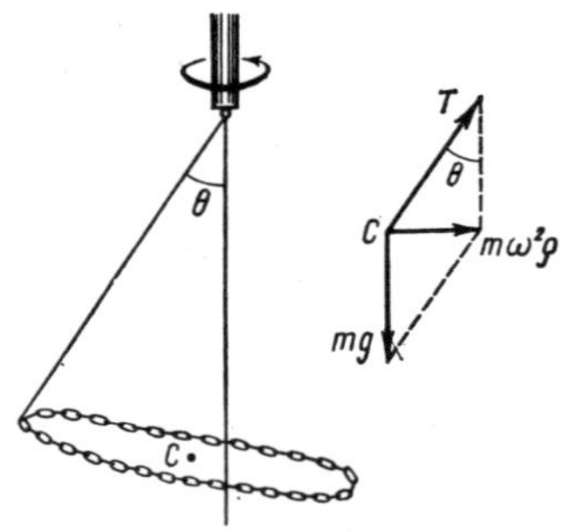

Fig. 58

The C frame. In many cases when we examine only the relative motion of particles within a system, but not the motion of this system as a whole, it is most advisable to resort to the reference frame in which the centre of inertia is at rest. Then we can significantly simplify both the analysis of phenomena and the calculations.

The reference frame rigidly fixed to the centre of inertia of a given system of particles and translating with respect to inertial frames is referred to as the *frame of the centre of inertia*, or, briefly, the *C frame*. The distinctive feature of the C frame is that the total momentum of the system of particles is equal to zero; this immediately follows from Eq. (4.10). In other words, any system of particles as a whole is at rest in its C frame.

The C frame of a closed system of particles is inertial, while that of a non-closed system is non-inertial in the general case.

Let an find the relationship between the values of the mechanical energy of a system in the K and C reference frames. Let us begin with the kinetic energy T of the system. The velocity of the ith particle in the K frame may be represented as $\mathbf{v}_i = \tilde{\mathbf{v}}_i + \mathbf{V}_C$, where $\tilde{\mathbf{v}}_i$ is the velocity of that particle in the C frame and $\mathbf{V}_C$ is the velocity of the C frame with respect to the K reference frame. Now we can write

$$T = \sum m_i v_i^2/2 = \sum m_i (\tilde{\mathbf{v}}_i + \mathbf{V}_C)^2/2 =$$
$$= \sum m_i \tilde{v}_i^2/2 + \mathbf{V}_C \sum m_i \tilde{\mathbf{v}}_i + \sum m_i \mathbf{V}_C^2/2.$$

Since in the C frame $\sum m_i \tilde{\mathbf{v}}_i = 0$, the previous expression takes the form

$$T = \tilde{T} + \frac{mV_C^2}{2} = \tilde{T} + \frac{p^2}{2m}, \tag{4.12}$$

where $\tilde{T} = \sum m_i \tilde{v}_i^2/2$ is the total kinetic energy of the particles in the C frame, m is the mass of all the system, and p is its total momentum in the K frame.

Thus, *the kinetic energy of a system of particles comprises the total kinetic energy $\tilde{T}$ in the C frame and the kinetic energy associated with the motion of the system of particles as a whole.* This important conclusion will be repeatedly utilized hereafter (specifically, in studies of dynamics of a solid).

It follows from Eq. (4.12) that the kinetic energy of a system of particles is minimal in the C frame, another distinctive feature of that frame. Indeed, in the C frame $\mathbf{V}_C = 0$, and Eq. (4.12) yields $T = \tilde{T}$.

Now let us pass over to the total mechanical energy E. Since the internal potential energy U of a system depends only on its configuration, the magnitude U is the same in all reference frames. Adding U to the left- and right-hand sides of Eq. (4.12), we obtain the formula for transformation of the total mechanical energy on transition from the K to the C frame:

$$E = \tilde{E} + \frac{mV_C^2}{2} = \tilde{E} + \frac{p^2}{2m}. \tag{4.13}$$

The energy $\tilde{E} = \tilde{T} + U$ is referred to as the *internal mechanical energy* of the system.

Example. Two small discs, each of mass m, lying on a smooth horizontal plane, are interconnected by a weightless spring. One of the discs is set in motion with the velocity v_0, as shown in Fig. 59. What is the internal mechanical energy $\tilde{E}$ of this system in the process of motion?

Since the surface is smooth, the system in the process of motion behaves as a closed one. Therefore, its total mechanical energy E and total momentum $\mathbf{p}$ remain constant and equal to the initial values, i.e. $E = mv_0^2/2$ and $p = mv_0$. Substituting these values into

Eq. (4.13), we obtain

$$\tilde{E} = E - p^2/(2\cdot 2m) = mv_0^2/4.$$

It is easy to realize that the internal energy $\tilde{E}$ is associated with the rotation and oscillation of the given system, while at the initial moment $\tilde{E}$ was equal only to the rotational motion energy.

If the processes associated with a change in the total mechanical energy take place in a *closed* system of particles, from Eq. (4.13) it follows that $\Delta E = \Delta\tilde{E}$, i.e. the increment of the total mechanical energy relative to an arbitrary inertial reference frame is equal to the increment of the *internal* mechanical energy. In this case the kinetic energy resulting from the motion of the system of particles as a whole does not change because in a closed system $\mathbf{p} = \text{const}$.

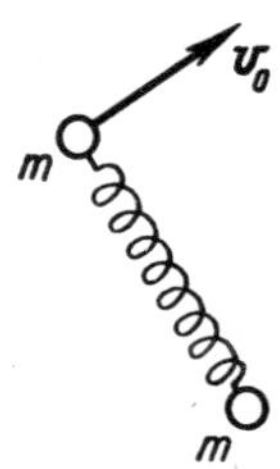

Fig. 59

Specifically, if a closed system is *conservative*, its total mechanical energy remains constant in all inertial reference frames. This conclusion completely agrees with the Galilean relativity principle.

A system of two particles. Suppose the masses of the particles are equal to m_1 and m_2 and their velocities in the K reference frame to $\mathbf{v}_1$ and $\mathbf{v}_2$, respectively. Let us find the expressions defining their momenta and the total kinetic energy in the C frame.

The momentum of the first particle in the C system is

$$\tilde{\mathbf{p}}_1 = m_1\tilde{\mathbf{v}}_1 = m_1(\mathbf{v}_1 - \mathbf{V}_C),$$

where $\mathbf{V}_C$ is the velocity of the centre of inertia (of the C system) in the K reference frame. Substituting in this formula expression (4.9) for $\mathbf{V}_C$, we obtain

$$\tilde{\mathbf{p}}_1 = \mu(\mathbf{v}_1 - \mathbf{v}_2),$$

where μ is the so-called *reduced mass* of the system

$$\boxed{\mu = m_1m_2/(m_1 + m_2).} \qquad (4.14)$$

Similarly, the momentum of the second particle in the C frame is

$$\tilde{\mathbf{p}}_2 = \mu(\mathbf{v}_2 - \mathbf{v}_1).$$

Thus, the momenta of the two particles in the C frame are equal in magnitude and opposite in direction; the modulus of the momentum of each particle is

$$\boxed{\tilde{p} = \mu v_{rel},} \qquad (4.15)$$

where $v_{rel} = |\mathbf{v}_1 - \mathbf{v}_2|$ is the velocity of one particle relative to another.

Finally, let us consider kinetic energy. The total kinetic energy of the two particles in the C frame is

$$\tilde{T} = \tilde{T}_1 + \tilde{T}_2 = \tilde{p}^2/2m_1 + \tilde{p}^2/2m_2.$$

Since in accordance with Eq. (4.14) $1/m_1 + 1/m_2 = 1/\mu$, then

$$\boxed{\tilde{T} = \tilde{p}^2/2\mu = \mu v_{rel}^2/2.} \qquad (4.16)$$

If the particles interact, their total mechanical energy in the C frame is

$$\tilde{E} = \tilde{T} + U, \qquad (4.17)$$

where U is the potential energy of interaction of the given particles.

The formulae obtained play an important part in studies of particle collisions.

§ 4.3. Collision of Two Particles

In this section we shall examine various cases of collisions of two particles, using only the momentum and energy conservation laws as an investigatory tool. Here we shall see that the conservation laws enable us to draw some general and essential conclusions concerning the properties of a given process irrespective of a specific law of particle interaction.

At the same time we shall illustrate the advantages of the C frame, whose utilization considerably simplifies analysis of a process and many calculations.

Although we shall discuss collisions of *particles*, it should be mentioned at once that all subsequent arguments and conclusions relate to collisions of *any bodies*. One only has to substitute the velocity of the centre of inertia of each body for the velocity of a particle, and to replace the kinetic energy of a particle by that part of the kinetic energy of each body that characterizes its motion as a whole.

In what follows we shall assume that

(1) the initial reference frame K is inertial,

(2) the system of two particles is closed,

(3) the momenta (and the velocities) of the particles before and after a collision correspond to sufficiently large distances between them; at the same time the potential energy of interaction can be neglected.

In addition, the quantities relating to the system after a collision will be marked with a prime, while those in the C frame with a tilde.

Now let us pass to the essence of the problem. Particle collisions are classified into three types: completely inelastic, perfectly elastic, and inelastic (the intermediate case).

Completely inelastic collision results in two particles "sticking together", after which they move as a single unit. Suppose two particles with masses m_1 and m_2 move with the velocities $\mathbf{v}_1$ and $\mathbf{v}_2$ before collision (in the K frame). After the collision a particle with mass $m_1 + m_2$ is formed because of additivity of mass in Newtonian mechanics. The velocity $\mathbf{v}'$ of the formed particle can be immediately found from the momentum conservation law:

$$(m_1 + m_2)\,\mathbf{v}' = m_1\mathbf{v}_1 + m_2\mathbf{v}_2.$$

The velocity $\mathbf{v}'$ is obviously equal to that of the system's centre of inertia.

In the C frame this process is the most simple: prior to the collision both particles move toward each other carrying equal momenta $\tilde{p}$, while after the collision the formed particle turns out to be stationary. In this case the total kinetic energy $\tilde{T}$ of the particles completely turns into the internal energy Q of the formed particle, i.e. $\tilde{T} = Q$. Whence, with allowance made for Eq. (4.16), we obtain

$$Q=\frac{\mu v_{rel}^2}{2}=\frac{1}{2}\frac{m_1 m_2}{m_1+m_2}(\mathbf{v}_1-\mathbf{v}_2)^2.$$

Thus the value of Q for a given pair of particles depends only on their relative velocity.

Perfectly elastic collision does not lead to any change in the internal energy of the particles, so that the kinetic energy of the system does not change either. We shall consider two particular cases: central (head-on) and non-central elastic collisions.

1. *A head-on collision.* Both particles move along the same straight line before and after collision. Suppose that prior to collision the particles move with the velocities $\mathbf{v}_1$

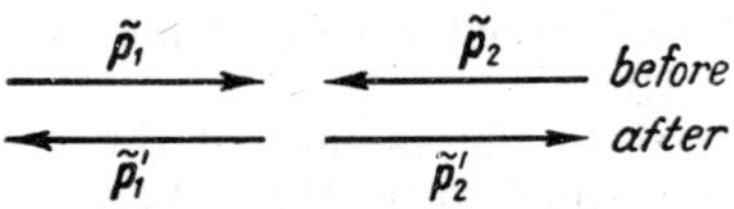

Fig. 60

and $\mathbf{v}_2$ in the K reference frame (the particles either move toward each other or one particle overtakes another). What are the velocities of these particles after the collision?

Let us first consider this process in the C frame, where the particles before and after the collision possess momenta equal in magnitude and opposite in direction (Fig. 60). Moreover, since the total kinetic energy of the particles is the same before and after the collision, as well as their reduced mass, then in accordance with Eq. (4.16) the momentum of each particle only reverses its direction as a result of the collision, its magnitude remaining unchanged, i.e. $\tilde{\mathbf{p}}_i' = -\tilde{\mathbf{p}}_i$, where $i = 1, 2$. The same can be said about the velocity of each particle in the C frame:

$$\tilde{\mathbf{v}}_i'=-\tilde{\mathbf{v}}_i.$$

Now let us find the velocity of each particle after the collision in the K reference frame. For this purpose we shall make use of the velocity transformation formulae for the transition from the C to the K frame and also the foregoing equality. Then

$$\mathbf{v}_i' = \mathbf{V}_C + \tilde{\mathbf{v}}_i' = \mathbf{V}_C - \tilde{\mathbf{v}}_i = \mathbf{V}_C - (\mathbf{v}_i - \mathbf{V}_C) = 2\mathbf{V}_C - \mathbf{v}_i,$$

where $\mathbf{V}_C$ is the velocity of the centre of inertia (of the C frame) in the K frame; this velocity is determined by Eq. (4.9). Hence, the velocity of the ith particle in the K frame after the collision is

$$\mathbf{v}_i' = 2\mathbf{V}_C - \mathbf{v}_i, \tag{4.18}$$

where $i = 1, 2$. In terms of the projections on an arbitrary x axis the last equality takes the form

$$v_{ix}' = 2V_{Cx} - v_{ix}. \tag{4.19}$$

Specifically, when the particle masses are identical, it is easy to see that the particles exchange their velocities as a result of the collision, i.e. $\mathbf{v}_1' = \mathbf{v}_2$ and $\mathbf{v}_2' = \mathbf{v}_1$.

2. *A non-central collision.* We shall limit ourselves to the case when one of the particles *is motionless* before the collision. Suppose a particle possessing the mass m_1 and momentum $\mathbf{p}_1$ experiences in the K frame a non-central elastic collision with a motionless particle of mass m_2. What are the possible momenta of these particles after the collision?

First, let us examine this process in the C frame. Here, as in the previous case, the particles possess momenta equal in magnitude and opposite in direction at any moment of time before and after the collision. Besides, the momentum of each particle does not change in magnitude following the collision, i.e.

$$\tilde{p}' = \tilde{p}.$$

However, the particles' rebound direction is different in this case. It forms a certain angle $\tilde{\theta}$ with the initial motion direction (Fig. 61), depending on the particle interaction law and the mutual positions of the particles in the process of collision.

Now let us calculate the momentum of each particle after the collision in the K reference frame. Making use of the velocity transformation formulae for the transition from the C to the K frame, we obtain:

$$\mathbf{p}_1' = m_1\mathbf{v}_1' = m_1(\mathbf{V}_C + \tilde{\mathbf{v}}_1') = m_1\mathbf{V}_C + \tilde{\mathbf{p}}_1',$$
$$\mathbf{p}_2' = m_2\mathbf{v}_2' = m_2(\mathbf{V}_C + \tilde{\mathbf{v}}_2') = m_2\mathbf{V}_C + \tilde{\mathbf{p}}_2', \qquad (4.20)$$

where $\mathbf{V}_C$ is the velocity of the C frame relative to the K reference frame.

Summing up separately the left- and right-hand sides of these equalities and taking into account that $\tilde{\mathbf{p}}_1' = -\tilde{\mathbf{p}}_2'$, we get

$$\mathbf{p}_1' + \mathbf{p}_2' = (m_1 + m_2)\mathbf{V}_C = \mathbf{p}_1,$$

just as it should be in accordance with the momentum conservation law.

Now let us draw the so-called *vector diagram of momenta*. First we depict the vector $\mathbf{p}_1$ as the section AB (Fig. 62),

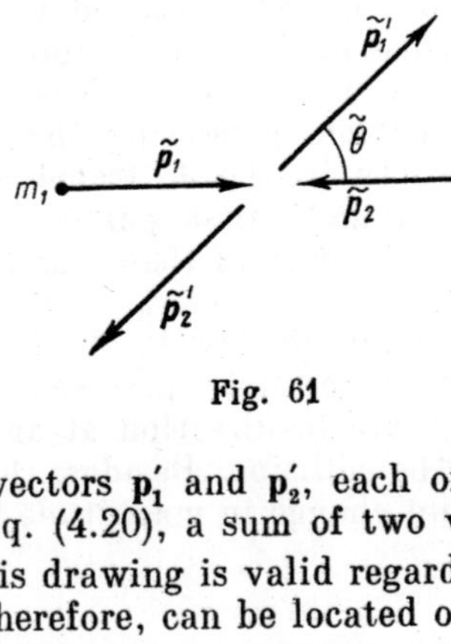

Fig. 61

and then the vectors $\mathbf{p}_1'$ and $\mathbf{p}_2'$, each of which represents, according to Eq. (4.20), a sum of two vectors.

Note that this drawing is valid regardless of the angle $\tilde{\theta}$. The point C, therefore, can be located only on the circle of radius $\tilde{p}$ having its centre at the point O, which divides the section AB into two parts in the ratio $AO : OB = m_1 : m_2$. Moreover, in the considered case (when the particle of mass m_2 rests prior to the collision) this circle passes through the point B, the end point of the vector $\mathbf{p}_1$, since the section $OB = \tilde{p}$. Indeed,

$$OB = m_2V_C = m_2\frac{m_1v_1}{m_1+m_2},$$

where v_1 is the velocity of the bombarding particle. But inasmuch as in our case $v_1 = v_{rel}$,

$$OB = \mu v_{rel} = \tilde{p}$$

in accordance with Eqs. (4.14) and (4.15). Thus, in order to draw a vector diagram of momenta corresponding to an

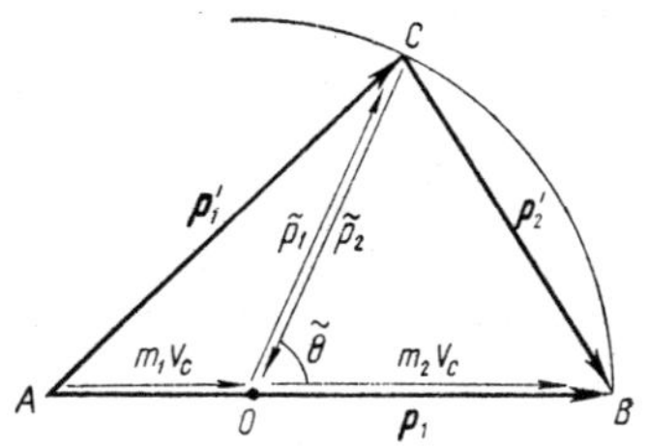

Fig. 62

elastic collision of two particles (one of which rests initially) it is necessary:

(1) first to depict the section AB equal to the momentum p_1 of the bombarding particle;

(2) then through the point B, the end point of the vector $\mathbf{p}_1$, to trace a circle of radius

$$\tilde{p} = \mu v_{rel} = \frac{m_2}{m_1 + m_2} p_1,$$

whose centre (point O) divides the section AB into two parts in the ratio $AO : OB = m_1 : m_2$.

This circle is the locus of all possible locations of the apex C of the momenta triangle ABC whose sides AC and CB represent the possible momenta of the particles after the collision (in the K reference frame).

Depending on the particle mass ratio the point A, the beginning of the vector $\mathbf{p}_1$, can be located inside the given circle, on it, or outside it (Fig. 63). In all three cases the angle $\tilde{\theta}$ can assume all the values from 0 to π. The possible values of the angle θ_1 of scattering of the bombarding particle and the angle Θ of rebounding particles are as follows:

(a) $m_1 < m_2$ $\quad 0 < \theta_1 \leqslant \pi$ $\quad \Theta > \pi/2$

(b) $m_1 = m_2$ $\quad 0 < \theta_1 \leqslant \pi/2$ $\quad \Theta = \pi/2$

(c) $m_1 > m_2$ $\quad 0 < \theta_1 \leqslant \theta_{1\,max}$ $\quad \Theta < \pi/2$

Here θ_{1max} is the *limiting angle*. It is defined by the formula

$$\sin \theta_{1\,max} = m_2/m_1, \tag{4.21}$$

which directly follows from Fig. 63*c*: $\sin \theta_{1max} = OC'/AO = OB/AO = m_2/m_1$.

In addition, here is another interesting fact. In the last case ($m_1 > m_2$) the particle m_1 can be scattered by the same

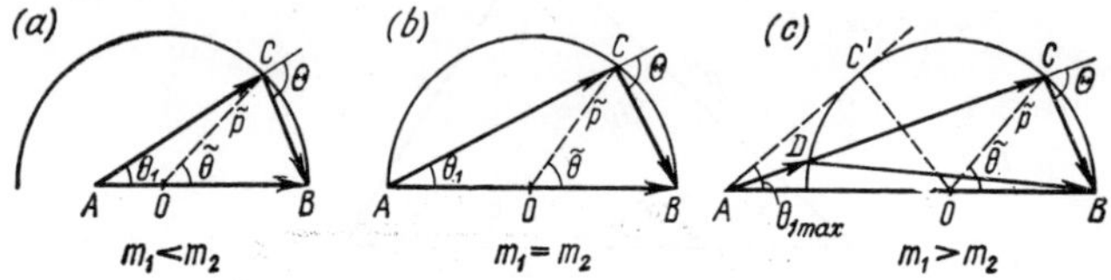

Fig. 63

angle θ_1 whether it possesses the momentum AC or AD (Fig. 63*c*), i.e. the solution is ambiguous. The same is true for the particle m_2.

And finally, from the same vector diagram of momenta we can determine the relation between the angles θ_1 and $\tilde{\theta}$

$$\tan \theta_1 = \frac{\sin \tilde{\theta}}{\cos \tilde{\theta} + m_1/m_2}. \tag{4.22}$$

With this we have exhausted any information on the given process which can be derived through the use of only the momentum and energy conservation laws.

Thus, we see that the momentum and energy conservation laws by themselves permit us to draw a number of significant conclusions about the properties of a given process. Most essential here is the fact that these properties are universal in their nature, that is, they do not depend on the type of interaction between particles.

One fundamental fact, however, should be pointed out. The vector diagram of momenta based on the momentum and energy conservation laws provides us with a complete pattern of all possible cases of rebounding particles; but this very significant result cannot indicate the *concrete* case that is actually realized. To answer that question, we must analyse the collision process in more detail by means of the motion equations. In the process it becomes clear that the scattering angle θ_1 of a bombarding particle depends on the type of the interaction of colliding particles and on the so-called *aiming parameter**, while the ambiguity of a solution in the case $m_1 > m_2$ is due to the fact that the same scattering angle θ_1 can occur with two different values of the aiming parameter irrespective of the law of particle interaction.

The circumstance discussed here represents an inherent, fundamental property of all conservation laws in general. The conservation laws can never provide an unambiguous picture of *what* is actually going to happen. But if, on the basis of some other considerations, it becomes possible to infer *what exactly* is going to happen, the conservation laws can contribute information on *how* it must happen.

Inelastic collision. After this kind of collision the internal energy of rebounding particles (or one of them) changes, and therefore the total kinetic energy of the system changes as well. It is customary to denote the corresponding increment of the kinetic energy of the system by Q. Depending on the sign of Q an inelastic collision is referred to as *exoergic* ($Q > 0$) or *endoergic* ($Q < 0$). In the former case the kinetic energy of the system increases while in the latter it decreases. In an elastic collision $Q = 0$, of course.

Our task is to determine possible momenta of particles after a collision.

This problem is easiest when solved in terms of the C frame. By the hypothesis, the increment of the total kinetic energy of the system in the given process is equal to

$$\widetilde{T}' - \widetilde{T} = Q. \qquad (4.23)$$

Since in this case $\widetilde{T}' \neq \widetilde{T}$, this means, in accordance with Eq. (4.16), that the momenta of particles change their magnitude after the collision. The momentum of each particle $\widetilde{p}'$ after the collision can be easily found if we replace

* The *aiming parameter* is the distance between the straight line along which the momentum of a bombarding particle is directed, and the particle exposed to a "collision".

$\widetilde{T}'$ in Eq. (4.23) by its expression $\widetilde{T}' = \widetilde{p}'^2/2\mu$:

$$\widetilde{p}' = \sqrt{2\mu(\widetilde{T}+Q)}. \tag{4.24}$$

Now let us consider the same problem in the K reference frame, where a particle of mass m_1 with the momentum p_1 collides with a *stationary* particle of mass m_2. To determine the possible cases of particle rebounding after the collision, it is helpful to resort to the vector diagram of momenta. It is drawn similarly to the case of an elastic collision. The momentum of a bombarding particle $p_1 = AB$ (Fig. 64) is divided by the point O into two parts proportional to the masses of the particles ($AO : OB = m_1 : m_2$). Then from the point O a circle is drawn with radius $\widetilde{p}'$ specified by Eq. (4.24). This circle is the locus of all possible positions of the vertex C of the triangle ABC whose sides AC and CB are equal to the momenta of the corresponding particles after the collision.

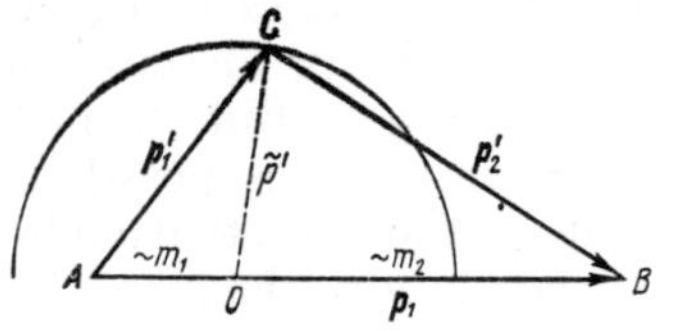

Fig. 64

Note that in contrast to the case of an elastic collision the point B, the end point of the vector $\mathbf{p}_1$, does not lie on the circle any more; in fact, when $Q>0$, this point is located inside the circle, and when $Q < 0$ outside it. Fig. 64 illustrates the latter case, an endoergic collision.

Threshold. There are many inelastic collisions in which the internal energy of particles is capable of changing by a quite *definite* value, depending on the properties of the particles themselves (e.g. inelastic collisions of atoms and molecules). Nevertheless, exoergic collisions ($Q > 0$) can occur for an arbitrarily low kinetic energy of a bombarding particle. In similar cases endoergic processes ($Q < 0$) possess a *threshold*. A threshold is the minimal kinetic energy of a bombarding particle just sufficient to make a given process possible in terms of energy.

So, suppose we need to carry out an endoergic collision in which the internal energy of the particles is capable of acquiring an increment not less than a certain value $|Q|$,

Under what condition does such a process become possible?

Again, the problem is easiest when solved in the C frame, where it is obvious that the total kinetic energy $\tilde{T}$ of the particles before the collision must in any case be not less than $|Q|$, i.e. $\tilde{T} \geqslant |Q|$. Whence it follows that there exists the minimal value $\tilde{T}_{min} = |Q|$, such that the kinetic energy of the system entirely turns into an increment of the internal energy of the particles, and so the particles *come to a standstill* in the C frame.

Let us consider the same problem in the K reference frame, where a particle of mass m_1 collides with a *stationary* particle of mass m_2. Since at $\tilde{T}_{min}$ the particles come to a standstill after the collision in the C frame, this signifies that in the K frame, provided the kinetic energy of the bombarding particle is equal to the requisite threshold value T_{1thr}, both particles move after the collision *as a single unit* whose total momentum is equal to the momentum $\mathbf{p}_1$ of the bombarding particle and the kinetic energy $p_1^2/2(m_1 + m_2)$. Therefore

$$T_{1\,thr} = |Q| + p_1^2/2(m_1 + m_2).$$

Taking into account that $T_{1thr} = p_1^2/2m_1$ and eliminating p_1^2 from these two equations, we obtain

$$T_{1\,thr} = \frac{m_1 + m_2}{m_2} |Q|. \tag{4.25}$$

This is the threshold kinetic energy of the bombarding particle sufficient to make the given endoergic process possible in terms of energy.

It should be pointed out that Eq. (4.25) plays an important part in atomic and nuclear physics. It is used to determine both the thresholds of various endoergic processes and their corresponding energies $|Q|$.

In conclusion we shall consider an example which, in essence, provides a model of an endoergic collision (see also Problems 4.5 and 4.8).

Example. A small disc of mass m and a smooth hillock of mass M and height h are located on a smooth horizontal plane (Fig. 65). What minimal velocity should be imparted to the disc to make it capable of overcoming the hillock?

It is clear that the velocity of the disc must be at least sufficient for it to climb the hillock and then to move together with it as a single unit. In the process, part of the system's kinetic energy turns into

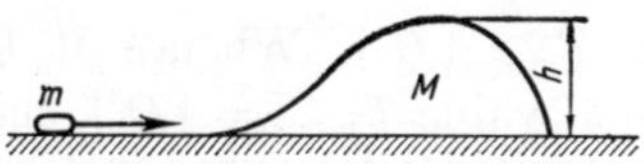

Fig. 65

an increment of potential energy $\Delta U = mgh$. We shall regard this process as endoergic in which $|Q| = \Delta U$. Then in accordance with Eq. (4.25).

$$mv_{thr}^2/2 = mgh\,(m+M)/M\,,$$

whence

$$v_{thr} = \sqrt{2\,(1+m/M)\,gh}.$$

§ 4.4. Motion of a Body with Variable Mass

There are many cases when the mass of a body varies in the process of motion due to the continuous separation or addition of matter (a missile, a jet, a flatcar being loaded in motion, etc.).

Our task is to derive the equation of motion of such a body.

Let us consider the solution of this problem for a mass point, calling it a body for the sake of brevity. Suppose that at a certain moment of time t the mass of a moving body A is equal to m and the mass being added (or separated) has the velocity $\mathbf{u}$ relative to the given body.

Let us introduce an auxiliary inertial reference frame K whose velocity is equal to that of the body A at a *given* moment t. This means that at the moment t the body A is at rest in the K frame.

Now suppose that during the time interval from t to $t + dt$ the body A acquires the momentum $m\,d\mathbf{v}$ in the K frame. The momentum is gathered due to (1) the addition (separation) of the mass δm bringing (carrying away) the momentum $\delta m \cdot \mathbf{u}$, and (2) surrounding bodies exerting the force $\mathbf{F}$, or the action of a field of force. Thus, it can be written that

$$m\,d\mathbf{v} = \mathbf{F}\,dt \pm \delta m \cdot \mathbf{u},$$

where the plus sign denotes the addition of mass and the minus sign denotes the separation. These cases can be combined, designating $\pm\delta m$ as the mass increment dm of the body A (in fact, in the case of mass addition $dm = +\delta m$, and in the case of mass separation $dm = -\delta m$). Then the foregoing equation takes the form

$$m\,d\mathbf{v} = \mathbf{F}\,dt + dm \cdot \mathbf{u}.$$

Dividing this expression by dt, we obtain

$$\boxed{m\frac{d\mathbf{v}}{dt} = \mathbf{F} + \frac{dm}{dt}\mathbf{u},} \tag{4.26}$$

where $\mathbf{u}$ is the velocity of the added (separated) matter with respect to the considered body.

This is the *fundamental equation of dynamics of a mass point with variable mass.* It is referred to as the *Meshchersky equation.* Obtained in one inertial reference frame, this equation is also valid, due to the relativity principle, in any other inertial frame. It should be pointed out that in a non-inertial reference frame the force $\mathbf{F}$ is interpreted as the resultant of both inertial forces and the forces of interaction of a given body with surrounding bodies.

The last term in Eq. (4.26) is referred to as the *reactive force*: $\mathbf{R} = (dm/dt)\,\mathbf{u}$. This force appears as a result of the action that the added (separated) mass exerts on a given body. If mass is added, then $dm/dt > 0$ and the vector $\mathbf{R}$ coincides in direction with the vector $\mathbf{u}$; if mass is separated, $dm/dt < 0$ and the vector $\mathbf{R}$ is directed oppositely to the vector $\mathbf{u}$.

The Meshchersky equation coincides in form with the fundamental equation of dynamics for a permanent mass point: the left-hand side contains the product of the mass of a body by acceleration, and the right-hand side contains the forces acting on it, including the reactive force. In the case of variable mass, however, we cannot include the mass m under the differential sign and present the left-hand side of the equation as the time derivative of the momentum, since $m\,d\mathbf{v}/dt \neq d\,(m\mathbf{v})/dt$.

Let us discuss two special cases.

1. When $\mathbf{u} = 0$, i.e. mass is added or separated with zero velocity relative to the body, $\mathbf{R} = 0$ and Eq. (4.26) takes the form

$$m(t)\frac{d\mathbf{v}}{dt} = \mathbf{F}, \tag{4.27}$$

where $m(t)$ is the mass of the body at a given moment of time. This equation describes, for example, the motion of a flatcar with sand pouring out freely from it (see Problem 4.10, Item 1).

2. If $\mathbf{u} = -\mathbf{v}$, i.e. the added mass is stationary in the chosen reference frame, or the separated mass becomes stationary in that frame, Eq. (4.26) takes another form,

$$m\,(d\mathbf{v}/dt) + (dm/dt)\,\mathbf{v} = \mathbf{F},$$

or

$$d\,(m\mathbf{v})/dt = \mathbf{F}. \tag{4.28}$$

In other words, in this case (and only in this one) the action of the force $\mathbf{F}$ determines the change of *momentum* of a body with variable mass. This is realized, for example, during the motion of a flatcar being loaded with sand from a stationary hopper (see Problem 4.10, Item 2).

Let us consider an example in which the Meshchersky equation is utilized.

Example. A rocket moves in the inertial reference frame K in the absence of an external field of force, the gaseous jet escaping with the constant velocity $\mathbf{u}$ relative to the rocket. Find how the rocket velocity depends on its mass m if at the moment of launching the mass is equal to m_0.

In this case $\mathbf{F} = 0$ and Eq. (4.26) yields

$$d\mathbf{v} = \mathbf{u}\,dm/m.$$

Integrating this expression with allowance made for the initial conditions, we get

$$\mathbf{v} = -\mathbf{u}\ln(m_0/m), \tag{1}$$

where the minus sign shows that the vector $\mathbf{v}$ (the rocket velocity) is directed oppositely to the vector $\mathbf{u}$. It is seen that in this case ($\mathbf{u}$ = const) the rocket velocity does not depend on the fuel combustion time: $\mathbf{v}$ is determined only by the ratio of the initial rocket mass m_0 to the remaining mass m.

Note that if the whole fuel mass were momentarily ejected with the velocity $\mathbf{u}$ relative to the rocket, the rocket velocity would be different. In fact, if the rocket initially is at rest in the chosen inertial reference frame and after the fuel ejection gathers the velocity **v**, the momentum conservation law for the system "rocket-and-fuel" yields

$$0 = m\mathbf{v} + (m_0 - m)(\mathbf{u} + \mathbf{v}),$$

where $\mathbf{u} + \mathbf{v}$ is the velocity of the fuel relative to the given reference frame. Hence

$$\mathbf{v} = -\mathbf{u}(1 - m/m_0). \tag{2}$$

In this case the rocket velocity **v** turns out to be less than in the previous case (for equal values of the m_0/m ratio). This is easy to demonstrate, having compared the dependence of **v** on m_0/m in both cases. In the first case (when matter separates continuously) the rocket velocity **v** grows infinitely with increasing m_0/m as Eq. (1) shows, while in the second case (when matter separates momentarily) the velocity **v** tends to the limiting value $-\mathbf{u}$ (see Eq. (2)).

Problems to Chapter 4

● 4.1. A particle moves with the momentum $\mathbf{p}(t)$ due to the force $\mathbf{F}(t)$. Let **a** and **b** be constant vectors, with $\mathbf{a} \perp \mathbf{b}$. Assuming that

(1) $\mathbf{p}(t) = \mathbf{a} + t(1 - \alpha t)\mathbf{b}$, where α is a positive constant, find the vector **F** at the moments of time when $\mathbf{F} \perp \mathbf{p}$;

(2) $\mathbf{F}(t) = \mathbf{a} + 2t\,\mathbf{b}$ and $\mathbf{p}(0) = \mathbf{p}_0$, where $\mathbf{p}_0$ is a vector directed oppositely to the vector **a**, find the vector **p** at the moment t_0 when it is turned through 90° with respect to the vector $\mathbf{p}_0$.

Solution. 1. The force $\mathbf{F} = d\mathbf{p}/dt = (1 - 2\alpha t)\,\mathbf{b}$, i.e. the vector **F** is always perpendicular to the vector **a**. Consequently, the vector **F** is perpendicular to the vector **p** at those moments when the coefficient of **b** in the expression for $\mathbf{p}(t)$ turns into zero. Hence, $t_1 = 0$ and $t_2 = 1/\alpha$; the respective values of the vector **F** are equal to

$$\mathbf{F}_1 = \mathbf{b}, \quad \mathbf{F}_2 = -\mathbf{b}.$$

2. The increment of the vector **p** during the time interval dt is $d\mathbf{p} = \mathbf{F}\,dt$. Integrating this equation with allowance made for the initial conditions, we obtain

$$\mathbf{p} - \mathbf{p}_0 = \int_0^t \mathbf{F}\,dt = \mathbf{a}t + \mathbf{b}t^2,$$

where by the hypothesis $\mathbf{p}_0$ is directed oppositely to the vector **a**. The vector **p** turns out to be perpendicular to the vector $\mathbf{p}_0$ at the moment t_0 when $at_0 = p_0$. At this moment $\mathbf{p} = (p_0/a)^2\,\mathbf{b}$.

● 4.2. A rope thrown over a pulley (Fig. 66) has a ladder with a man *A* on one of its ends and a counterbalancing mass *M* on its other end. The man, whose mass is *m*, climbs upward by $\Delta\mathbf{r}'$ relative to the ladder and then stops. Ignoring the masses of the pulley and the rope, as well as the friction in the pulley axis, find the displacement of the centre of inertia of this system.

Solution. All the bodies of the system are initially at rest, and therefore the increments of momenta of the bodies in their motion are equal to the momenta themselves. The rope tension is the same both on the left- and on the right-hand side, and consequently the momenta of the counterbalancing mass ($\mathbf{p}_1$) and the ladder with the man ($\mathbf{p}_2$) are equal at any moment of time, i.e. $\mathbf{p}_1 = \mathbf{p}_2$, or

$$M\mathbf{v}_1 = m\mathbf{v} + (M - m)\,\mathbf{v}_2,$$

where $\mathbf{v}_1$, $\mathbf{v}$, and $\mathbf{v}_2$ are the velocities of the mass, the man, and the ladder, respectively. Taking into account that $\mathbf{v}_2 = -\mathbf{v}_1$ and $\mathbf{v} = \mathbf{v}_2 + \mathbf{v}'$, where $\mathbf{v}'$ is the man's velocity relative to the ladder, we obtain

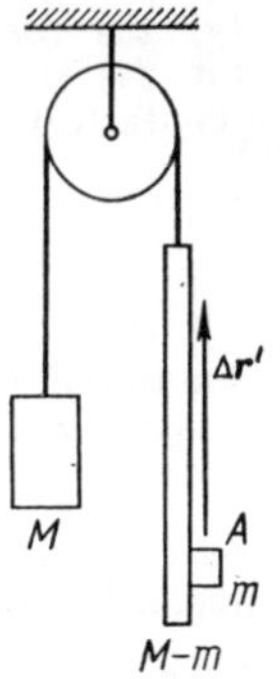

Fig. 66

$$\mathbf{v}_1 = (m/2M)\,\mathbf{v}'. \qquad (1)$$

On the other hand, the momentum of the whole system

$$\mathbf{p} = \mathbf{p}_1 + \mathbf{p}_2 = 2\mathbf{p}_1, \quad \text{or} \quad 2M\,\mathbf{V}_C = 2M\mathbf{v}_1,$$

where $\mathbf{V}_C$ is the velocity of the centre of inertia of the system. With allowance made for Eq. (1) we get

$$\mathbf{V}_C = \mathbf{v}_1 = (m/2M)\,\mathbf{v}'.$$

And finally, the sought displacement is

$$\Delta\mathbf{r}_C = \int \mathbf{V}_C\,dt = (m/2M)\int \mathbf{v}'\,dt = (m/2M)\,\Delta\mathbf{r}'.$$

Another method of solution is based on a property of the centre of inertia. In the reference frame fixed to the pulley axis the location of the centre of inertia of the given system is described by the radius vector

$$\mathbf{r}_C = [M\mathbf{r}_1 + (M - m)\,\mathbf{r}_2 + m\mathbf{r}_3]/2M,$$

where $\mathbf{r}_1$, $\mathbf{r}_2$, and $\mathbf{r}_3$ are the radius vectors of the centres of inertia of the mass M, the ladder, and the man relative to some point O of the given reference frame. Hence, the displacement of the centre of inertia $\Delta\mathbf{r}_C$ is equal to

$$\Delta\mathbf{r}_C = [M\,\Delta\mathbf{r}_1 + (M - m)\,\Delta\mathbf{r}_2 + m\,\Delta\mathbf{r}_3]/2M,$$

where $\Delta\mathbf{r}_1$, $\Delta\mathbf{r}_2$, and $\Delta\mathbf{r}_3$ are the displacements of the mass M, the ladder, and the man relative to the given reference frame. Since $\Delta\mathbf{r}_1 = -\Delta\mathbf{r}_2$ and $\Delta\mathbf{r}_3 = \Delta\mathbf{r}_2 + \Delta\mathbf{r}'$, we obtain

$$\Delta\mathbf{r}_C = (m/2M)\,\Delta\mathbf{r}'.$$

● 4.3. A system comprises two small spheres with masses m_1 and m_2 interconnected by a weightless spring. The spheres are set in motion at the velocities $\mathbf{v}_1$ and $\mathbf{v}_2$, as shown in Fig. 67, whereupon the system starts moving in the uniform gravitational field of the Earth. Ignoring the air drag and assuming that the spring is non-deformed at the initial moment of time, find:

(1) the velocity $\mathbf{V}_C(t)$ of the centre of inertia of this system as a function of time;

(2) the internal mechanical energy of the system in the process of motion.

Solution. 1. In accordance with Eq. (4.11) the velocity vector increment of the centre of inertia is $d\,\mathbf{V}_C = \mathbf{g}\,dt$. Integrating this equation, we get $\mathbf{V}_C(t) - \mathbf{V}_C(0) = \mathbf{g}t$, where $\mathbf{V}_C(0)$ is the initial velocity of the centre of inertia. Hence

$$\mathbf{V}_C(t) = (m_1\mathbf{v}_1 + m_2\mathbf{v}_2)/(m_1 + m_2) + \mathbf{g}t.$$

2. The internal mechanical energy of a system is its energy $\widetilde{E}$ in the C frame. In this case the C frame moves with the acceleration $\mathbf{g}$, so that each sphere experiences two external forces in that frame: gravity $m_i\mathbf{g}$ and the inertial force $-m_i\mathbf{g}$. The total work performed

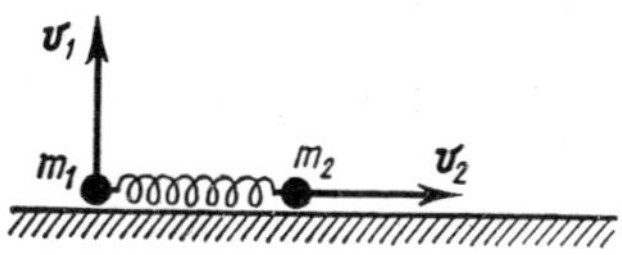

Fig. 67

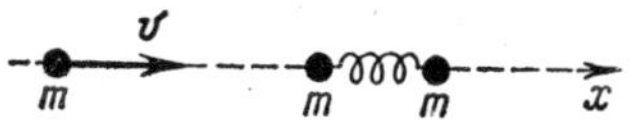

Fig. 68

by the external forces is thus equal to zero (in the C frame), and therefore the energy $\widetilde{E}$ does not change. To find the energy, it is sufficient to consider the initial moment of time, when the spring is not yet deformed and the energy $\widetilde{E}$ is equal to the kinetic energy $\widetilde{T}_0$ in the C frame. Making use of Eq. (4.16), we get

$$\widetilde{E} = \widetilde{T}_0 = \frac{\mu}{2}(\mathbf{v}_1 - \mathbf{v}_2)^2 = \frac{m_1 m_2}{2(m_1 + m_2)}(v_1^2 + v_2^2).$$

● **4.4.** A ball possessing the kinetic energy T collides head-on with an initially stationary elastic dumbbell (Fig. 68) and rebounds in the opposite direction with the kinetic energy T'. The masses of all three balls are the same. Find the energy of the dumbbell oscillations after the collision.

Solution. Suppose p and p' are the momenta of the striking ball before and after the collision, p'_C and T'_C are the momentum and the kinetic energy of the dumbbell as a whole after the collision, and E is the oscillation energy. In accordance with the momentum and energy conservation laws

$$p = -p' + p'_C, \qquad T = T' + T'_C + E.$$

Taking into account that $T = p^2/2m$, we obtain from these two equations:

$$E = (T - 3T' - 2\sqrt{TT'})/2.$$

● 4.5. In the K frame particle *1* of mass m_1 strikes a stationary particle *2* of mass m_2. The charge of each particle is equal to $+q$. Find the minimal distance separating the particles during a head-on collision if the kinetic energy of particle *1* is equal to T_1 when it is far removed from particle *2*.

Solution. Let us consider this process both in the K frame and in the C frame.

1. In the K frame the particles move at the moment of the closest approach as a single unit with the velocity v, which can be determined from the momentum conservation law:

$$p_1 = (m_1 + m_2)\, v,$$

where p_1 is the momentum of the striking particle, $p_1 = \sqrt{2m_1T_1}$. On the other hand, it follows from the energy conservation law that

$$T_1 = (m_1 + m_2)\, v^2/2 + \Delta U,$$

where the increment of the system's potential energy $\Delta U = kq^2/r_{min}$. Eliminating v from these two equations, we get

$$r_{min} = (kq^2/T_1)(1 + m_1/m_2).$$

2. The solution is simplest in the C frame: here the total kinetic energy of the particles turns entirely into an increment of the potential energy of the system at the moment of the closest approach:

$$\widetilde{T} = \Delta U,$$

where in accordance with Eq. (4.16) $\widetilde{T} = \mu v_1^2/2 = T_1 m_2/(m_1 + m_2)$, $\Delta U = kq^2/r_{min}$. From this it is easy to find r_{min}.

● 4.6. A particle of mass m_1 and momentum p_1 collides elastically with a stationary particle of mass m_2. Find the momentum p_1' of the first particle after its collision and scattering through the angle θ relative to the initial motion direction.

Solution. From the momentum conservation law we find (Fig. 69)

$$p_2'^2 = p_1^2 + p_1'^2 - 2p_1p_1' \cos\theta, \qquad (1)$$

where p_2' is the momentum of the second particle after the collision.

On the other hand, from the energy conservation law it follows that $T_1 = T_1' + T_2'$, where T_1' and T_2' are the kinetic energies of the first and second particles after the collision. Using the relation $T = p^2/2m$, we can reduce Eq. (1) to the following form:

$$p_2'^2 = (p_1^2 - p_1'^2)\, m_2/m_1. \qquad (2)$$

Eliminating $p_2'^2$ from Eqs. (1) and (2), we obtain

$$p_1' = p_1 \frac{\cos\theta \pm \sqrt{\cos^2\theta + (m_2^2/m_1^2 - 1)}}{1 + m_2/m_1}.$$

When $m_1 < m_2$, only the plus sign (in front of the radical sign) has physical meaning. This follows from the fact that under this condition the radical is greater than cos θ and that p_1' is the vector's modulus, which cannot be negative.

But when $m_1 > m_2$, both signs have physical meaning; the solution is ambiguous in this case: the momentum of the particle scattered through the angle θ may have one of two values (depending on the relative positioning of the particles at the moment of collision). The latter case is illustrated by the vector diagram shown in Fig. 63c.

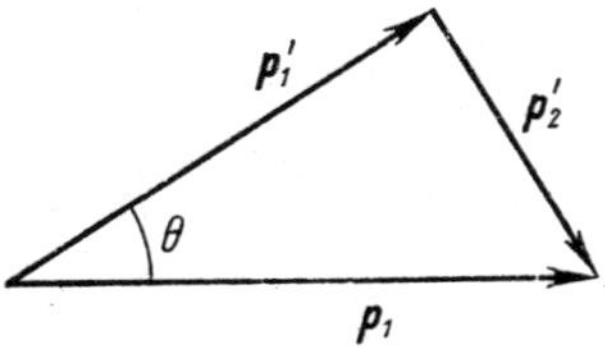

Fig. 69

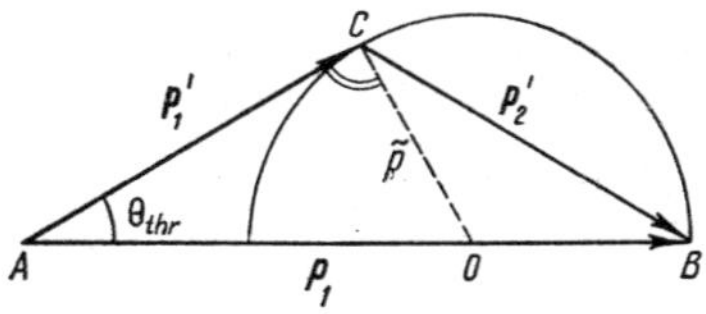

Fig. 70

●4.7. What fraction η of its kinetic energy does a particle of mass m_1 lose when it scatters, after an elastic collision, through the threshold angle on a stationary particle of mass m_2 ($m_1 > m_2$)?

Solution. Suppose T_1, p_1, T_1', and p_1' are the values of the kinetic energy and momentum of the striking particle before and after the scattering, respectively; then

$$\eta = (T_1 - T_1')/T_1 = 1 - T_1'/T_1 = 1 - (p_1'/p_1)^2, \quad (1)$$

i.e. the problem reduces to the determination of p_1'/p_1.

Let us make use of the vector diagram of momenta corresponding to the threshold angle θ_{thr} (Fig. 70). From the right triangle ACO it follows that

$$p_1'^2 = (p_1 - \tilde{p})^2 - \tilde{p}^2 = p_1^2 - 2p_1\tilde{p},$$

whence

$$(p_1'/p_1)^2 = 1 - 2\tilde{p}/p_1 = 1 - 2m_2/(m_1 + m_2). \quad (2)$$

Substituting Eq. (2) into Eq. (1), we obtain

$$\eta = 2m_2/(m_1 + m_2).$$

● 4.8. An atom of mass m_1 collides inelastically with a stationary molecule of mass m_2. After the collision the particles rebound at the angle Θ between them with kinetic energies T_1' and T_2' respectively. In the process the molecule is excited, that is, its internal energy increases by the definite value Q. Find Q and the threshold kinetic energy of the atom enabling the molecule to pass into the given excited state.

Solution. From the energy and momentum conservation laws in this process we can write

$$T_1 = T_1' + T_2' + Q,$$

$$p_1^2 = p_1'^2 + p_2'^2 + 2p_1'p_2' \cos\Theta,$$

where the primes mark the after-collision values. The second relation follows immediately from the momenta triangle according to the cosine theorem. Making use of the formula $p^2 = 2mT$, we eliminate T_1 from these equations and get

$$Q = (m_2/m_1 - 1)\, T_2' + 2\sqrt{(m_2/m_1)\, T_1' T_2'} \cos\Theta$$

and

$$T_{1\,thr} = |Q|\, (m_1 + m_2)/m_2.$$

●4.9. A particle with the momentum $\mathbf{p}_0$ (in the K frame) disintegrates into two particles with masses m_1 and m_2. The disintegration energy Q turns into kinetic energy. Draw the vector diagram of momenta for this process to find all possible momenta $\mathbf{p}_1$ and $\mathbf{p}_2$ of the generated particles.

Solution. This process appears to be the simplest in the C frame: here the disintegrating particle is at rest while the generated particles

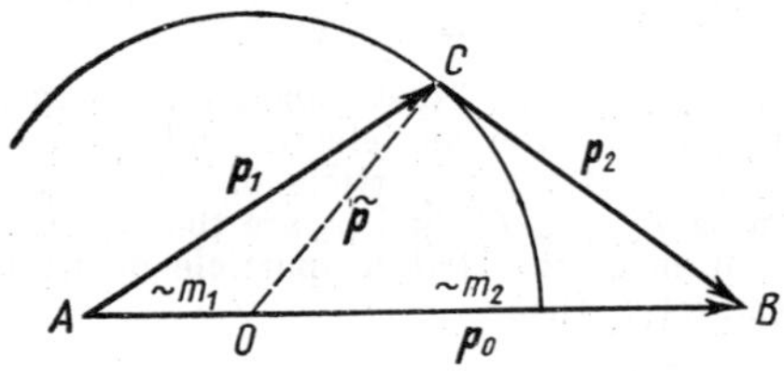

Fig. 71

move in opposite directions with momenta equal in magnitude $\tilde{p}_1 = \tilde{p}_2 = \tilde{p}$. Since the disintegration energy Q turns entirely into the total kinetic energy $\tilde{T}$ of the generated particles,

$$\tilde{p} = \sqrt{2\mu\tilde{T}} = \sqrt{2\mu Q},$$

where μ is the reduced mass of the generated particles.

Now let us find the momenta of these particles in the K frame. Making use of the velocity transformation formula for the transition from the C to the K frame, we can write:

$$\mathbf{p}_1 = m_1\mathbf{v}_1 = m_1(\mathbf{V}_C + \tilde{\mathbf{v}}_1) = m_1\mathbf{V}_C + \tilde{\mathbf{p}}_1,$$

$$\mathbf{p}_2 = m_2\mathbf{v}_2 = m_2(\mathbf{V}_C + \tilde{\mathbf{v}}_2) = m_2\mathbf{V}_C + \tilde{\mathbf{p}}_2$$

with $\mathbf{p}_1 + \mathbf{p}_2 = \mathbf{p}_0$ in accordance with the momentum conservation law.

Using these formulae, we can draw the vector diagram of momenta (Fig. 71). First, we draw the segment AB equal to the momentum $\mathbf{p}_0$. Then we draw a circle of radius $\tilde{p}$ from the point O to divide the segment AB into two parts in the ratio $m_1 : m_2$. This circle is the locus of all possible positions of the vertex C of the momenta triangle ABC.

● **4.10.** A flatcar starts rolling at the moment $t = 0$ due to the permanent force $\mathbf{F}$. Ignoring friction in the axes, find the time dependence of the velocity of the flatcar if

(1) it is loaded with sand which pours out through a hole in the bottom at the constant rate μ kg/s, and the initial mass of the flatcar with sand at the moment $t = 0$ is equal to m_0;

(2) the sand is loaded on it from a stationary hopper at the permanent rate μ kg/s, starting from the moment $t = 0$, when it had the mass m_0.

Solution. 1. In this case the reactive force is equal to zero and the Meshchersky equation (4.26) takes the form $(m_0 - \mu t)\, d\mathbf{v}/dt = \mathbf{F}$, whence

$$d\mathbf{v} = \mathbf{F}\, dt/(m_0 - \mu t).$$

Integrating this expression with allowance made for the initial conditions, we get

$$\mathbf{v} = (\mathbf{F}/\mu) \ln [m_0/(m_0 - \mu t)].$$

2. In this case the horizontal component of the reactive force (which is the only one of interest here) is $\mathbf{R} = \mu\,(-\mathbf{v})$, where $\mathbf{v}$ is the velocity of the flatcar. That is why the Meshchersky equation should be taken in the form (4.28), or

$$d\,(m\mathbf{v}) = \mathbf{F}\, dt.$$

Integrating this equation with allowance made for the initial conditions, we obtain

$$m\mathbf{v} = \mathbf{F}t,$$

where $m = m_0 + \mu t$. Hence,

$$\mathbf{v} = \mathbf{F}t/(m_0 + \mu t).$$

Needless to say, the expressions obtained in both cases are valid only in the process of unloading (or loading) a flatcar.

● **4.11.** A spaceship of mass m_0 moves with the constant velocity $\mathbf{v}_0$ in the absence of a field of force. To change the direction of motion, a reactive engine is started whose jet moves with the constant velocity $\mathbf{u}$ with respect to the spaceship and is directed perpendicular to the spaceship's direction of motion. The engine stops when the spaceship's mass is equal to m. Find how much the course of the spaceship changes during the operation of the engine.

Solution. Let us find the increment of the spaceship's velocity vector during the time interval dt. Multiplying both sides of the Meshchersky equation (4.26) by dt and taking into account that $\mathbf{F} = 0$, we get

$$d\mathbf{v} = \mathbf{u}\, dm/m.$$

Here $dm < 0$. Since the vector **u** is always perpendicular to the vector **v** (the spaceship's velocity), the modulus of the vector **v** does not change and retains its original magnitude: $|\mathbf{v}| = |\mathbf{v}_0| = v_0$. It follows from this that the rotation angle $d\alpha$ of the vector **v** that occurs during the time interval dt is equal to

$$d\alpha = |d\mathbf{v}|/v_0 = (u/v_0)\,|dm/m|.$$

Integrating this equation, we obtain

$$\alpha = (u/v_0) \ln (m_0/m).$$

CHAPTER 5

THE LAW OF CONSERVATION OF ANGULAR MOMENTUM

§ 5.1. Angular Momentum of a Particle. Moment of Force

The analysis of the behaviour of systems indicates that apart from energy and momentum there is still another mechanical quantity also associated with a conservation law, the so-called *angular momentum**. What is this quantity and what are its properties?

First, let us consider one particle. Suppose $\mathbf{r}$ is the radius vector describing its position relative to some point O of a chosen reference frame and $\mathbf{p}$ is its momentum in that frame. The angular momentum of the particle A relative to the point O (Fig. 72) is the vector $\mathbf{L}$ equal to the vector product of the vectors $\mathbf{r}$ and $\mathbf{p}$:

$$\boxed{\mathbf{L}=[\mathbf{rp}].} \tag{5.1}$$

It follows from this definition that $\mathbf{L}$ is an axial vector. Its direction is chosen so that the rotation about the point O toward the vector $\mathbf{p}$ and the vector $\mathbf{L}$ correspond to a right-handed screw. The modulus of the vector $\mathbf{L}$ is equal to

$$L = rp \sin \alpha = lp, \tag{5.2}$$

where α is the angle between $\mathbf{r}$ and $\mathbf{p}$, and $l = r \sin \alpha$ is the arm of the vector $\mathbf{p}$ relative to the point O (Fig. 72).

The equation of moments. Let us determine what mechanical quantity is responsible for the variation of the vector $\mathbf{L}$ in a given reference frame. For this purpose we differentiate Eq. (5.1) with respect to time:

$$d\mathbf{L}/dt = [d\mathbf{r}/dt,\ \mathbf{p}] + [\mathbf{r},\ d\mathbf{p}/dt].$$

Since the point O is stationary, the vector $d\mathbf{r}/dt$ is equal to the velocity $\mathbf{v}$ of the particle, i.e. coincides in its direction with the vector $\mathbf{p}$; therefore

$$[d\mathbf{r}/dt,\ \mathbf{p}] = 0.$$

---

* The following names are also used: *moment of momentum*, *moment of quantity of motion*, *rotational moment*, or simply *moment*.

Next, in accordance with Newton's second law $d\mathbf{p}/dt = \mathbf{F}$, where $\mathbf{F}$ is the resultant of all the forces applied to the particle. Consequently,

$$d\mathbf{L}/dt = [\mathbf{rF}].$$

The quantity on the right-hand side of this equation is referred to as the *moment of force*, or torque, of $\mathbf{F}$ relative

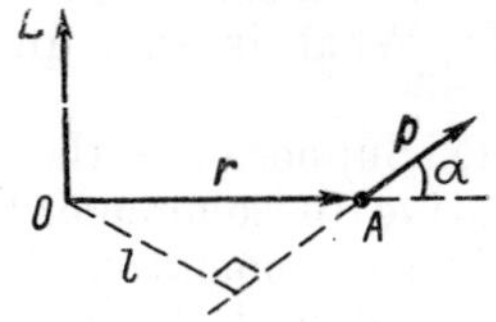

Fig. 72

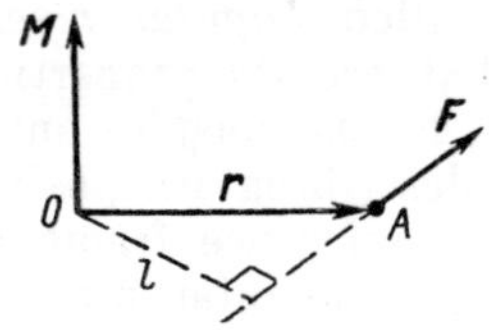

Fig. 73

to the point O (Fig. 73). Denoting it by the letter $\mathbf{M}$, we write

$$\boxed{\mathbf{M} = [\mathbf{rF}].} \tag{5.3}$$

The vector $\mathbf{M}$, like the vector $\mathbf{L}$, is axial. Similarly to (5.2) the modulus of this vector is equal to

$$M = lF, \tag{5.4}$$

where l is the arm of the vector $\mathbf{F}$ relative to the point O (Fig. 73).

Thus, the time derivative of the angular momentum $\mathbf{L}$ of the particle relative to some point O of the chosen reference frame is equal to the moment $\mathbf{M}$ of the resultant force $\mathbf{F}$ relative to the same point O:

$$\boxed{d\mathbf{L}/dt = \mathbf{M}.} \tag{5.5}$$

This equation is referred to as the *equation of moments*. Note that in the case of a noninertial reference frame the moment of the force $\mathbf{M}$ includes both the moment of the interaction forces and the moment of inertial forces (relative to the same point O).

Among other things, from the equation of moments (5.5) it follows that if $\mathbf{M} \equiv 0$, then $\mathbf{L} = \text{const}$. In other words, if the moment of all the forces acting on a particle relative to a certain point O of a chosen reference frame is equal to zero during the time interval of interest to us, the angular momentum of the particle relative to this point remains constant during that time.

Example 1. A planet A moves in the gravitational field of the Sun S (Fig. 74). Find the point in the heliocentric reference frame relative to which the angular momentum of that planet does not change in the course of time.

First of all, let us define what forces act on the planet A. In the given case it is only the gravitational force **F** of the Sun. Since during the motion of the planet the direction of this force passes through

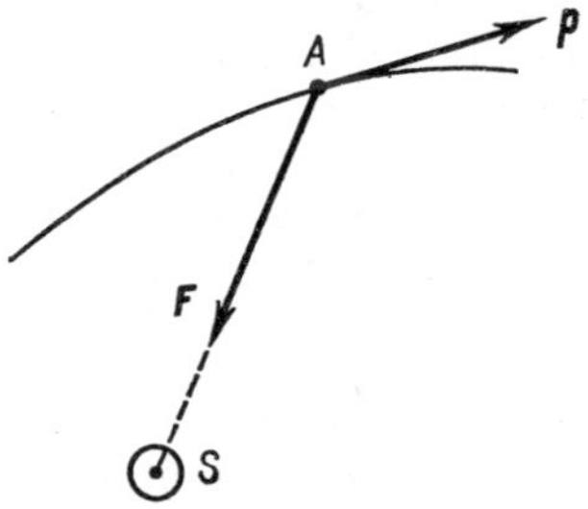

Fig. 74

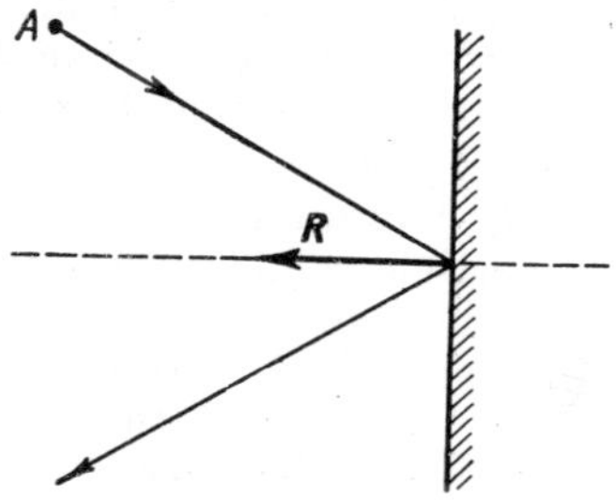

Fig. 75

the centre of the Sun, that point is the one relative to which the moment of the force is equal to zeroa nd the angular momentum of the planet remains constant. The momentum **p** of the planet changes in the process.

Example 2. A disc A moving on a smooth horizontal plane rebounds elastically from a smooth vertical wall (Fig. 75, top view). Find the point relative to which the angular momentum of the disc remains constant in this process.

The disc experiences gravity, the force of reaction of the horizontal surface, and the force **R** of reaction of the wall at the moment of the impact against it. The first two forces counterbalance each other, leaving only the force **R**. Its moment relative to any point of the line along which the vector **R** acts is equal to zero, and therefore the angular momentum of the disc relative to any of these points does not change in the given process.

Example 3. On a horizontal smooth plane there are a motionless vertical cylinder and a disc A connected to the cylinder by a thread AB (Fig. 76, top view). The disc is set in motion with the initial velocity **v** as shown in the figure. Is there any point here relative to which the angular momentum of the disc is invariable in the process of motion?

The only uncompensated force acting on the disc A in this case is the tension $\mathbf{F}$ of the thread. It is easy to see that there is no point here relative to which the moment of the force $\mathbf{F}$ is invariable in the process of motion. Therefore, there is *no* point relative to which the angular momentum of the disc would vary.

This example illustrates that sometimes a point relative to which the angular momentum of a particle is constant cannot be found at all.

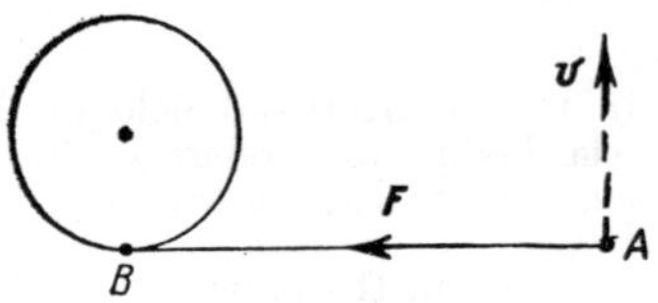

Fig. 76

The equation of momenta (5.5) makes it possible to solve the following two problems:

(1) find the force moment $\mathbf{M}$ relative to a certain point O at any moment of time t if the time dependence of the angular momentum $\mathbf{L}(t)$ of a particle relative to the same point is known;

(2) determine the increment of the angular momentum of a particle relative to a point O at any moment of time if the time dependence of the force moment $\mathbf{M}(t)$ acting on this particle (relative to the same point) is known.

The solution of the first problem reduces to the calculation of the time derivative of the angular momentum, that is, $d\mathbf{L}/dt$, which is equal, in accordance with Eq. (5.5), to the sought force moment $\mathbf{M}$.

The second problem is solved by integrating Eq. (5.5). Multiplying both sides of this equation by dt, we obtain the expression $d\mathbf{L} = \mathbf{M}\, dt$ determining the increment of the vector $\mathbf{L}$. Integrating this expression with respect to time, we get the increment of the vector $\mathbf{L}$ over the finite time interval t:

$$\mathbf{L}_2 - \mathbf{L}_1 = \int_0^t \mathbf{M}\, dt. \tag{5.6}$$

The quantity on the right-hand side of this equation is referred to as the *momentum of the force moment*, or the *torque momentum*. Thus, the increment of the angular momentum of a particle during any time interval is equal to the momentum of the force moment during the same time.

Let us consider two examples.

Example 1. The angular momentum of a particle relative to a certain point varies in the course of time t as $\mathbf{L}(t) = \mathbf{a} + \mathbf{b}t^2$, where $\mathbf{a}$ and $\mathbf{b}$ are certain constant vectors, with $\mathbf{a} \perp \mathbf{b}$. Find the force moment $\mathbf{M}$ acting on the particle when the angle between the vectors $\mathbf{M}$ and $\mathbf{L}$ is equal to 45°.

In accordance with Eq. (5.5) $\mathbf{M} = d\mathbf{L}/dt = 2\mathbf{b}t$, i.e. the vector $\mathbf{M}$ coincides in its direction with the vector $\mathbf{b}$. Let us depict the vectors $\mathbf{M}$

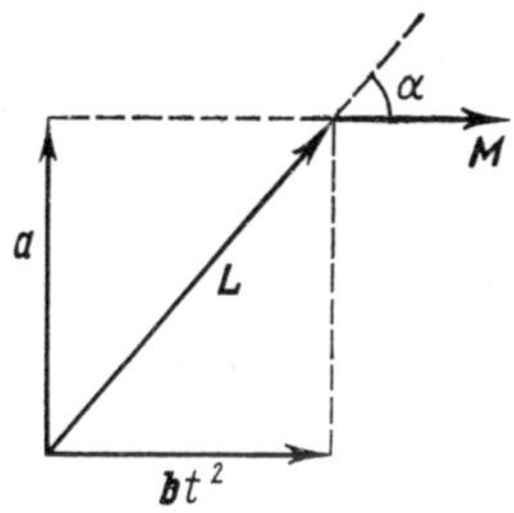

Fig. 77

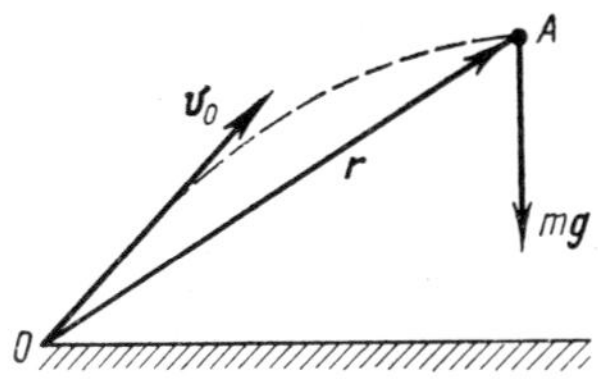

Fig. 78

and $\mathbf{L}$ at some moment t (Fig. 77). It is seen from the figure that the angle $\alpha = 45°$ at the moment t_0, when $a = bt_0^2$. Hence, $t_0 = \sqrt{a/b}$ and $\mathbf{M} = 2\sqrt{a/b}\,\mathbf{b}$.

Example 2. A stone A of mass m is thrown at an angle to the horizontal with the initial velocity $\mathbf{v}_0$. Ignoring the air drag, find the time dependence $\mathbf{L}(t)$ of the angular momentum of the stone relative to the point O from which the stone was thrown (Fig. 78).

During the time interval dt the angular momentum of the stone relative to the point O increases by $d\mathbf{L} = \mathbf{M}\,dt = [\mathbf{r}, m\mathbf{g}]\,dt$. Since $\mathbf{r} = \mathbf{v}_0 t + \mathbf{g}t^2/2$ (see p. 17), $d\mathbf{L} = [\mathbf{v}_0, m\mathbf{g}]\,t\,dt$. Integrating this expression with allowance made for the initial condition ($\mathbf{L}(0) = 0$ at $t = 0$), we get $\mathbf{L}(t) = [\mathbf{v}_0, m\mathbf{g}]\,t^2/2$. It is seen from this that the direction of the vector $\mathbf{L}$ remains constant in the process of motion (the vector $\mathbf{L}$ is directed beyond the plane of Fig. 78).

The angular momentum and the force moment relative to an axis. Let us choose an arbitrary motionless axis z in a given reference frame. Suppose the angular momentum of the particle A relative to a certain point O of the z axis is equal to $\mathbf{L}$ and the force moment acting on the particle is equal to $\mathbf{M}$.

The angular momentum relative to the z axis is the projection of the vector $\mathbf{L}$, defined with respect to an arbitrary point O of the given axis, on that axis (Fig. 79). The concept of a force moment relative to an axis is introduced in a similar fashion. They are denoted by L_z and M_z respectively. We shall see later that L_z and M_z do not depend on the choice of the point O on the z axis.

Let us examine the properties of these quantities. Projecting Eq. (5.5) on the z axis, we obtain

$$dL_z/dt = M_z, \tag{5.7}$$

i.e. the time derivative of the angular momentum of the particle relative to the z axis is equal to the force moment

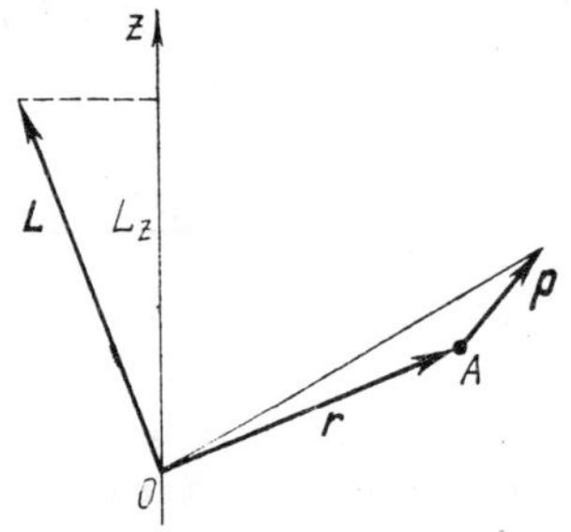

Fig. 79

relative to the same axis. In particular, if $M_z \equiv 0$, then $L_z = \text{const}$. In other words, when the force moment relative to a fixed axis z is equal to zero, the angular momentum of the particle relative to that axis remains constant. The vector **L** can, however, vary in the process.

Example. A small body of mass m suspended on a thread moves uniformly along a horizontal circle (Fig. 80) due to gravity $m\mathbf{g}$ and the tension **T** of the thread. The angular momentum of the body relative to the point O, the vector **L**, is located in the same plane as the z axis and the thread. During the motion of the body the vector **L** rotates continuously under the action of the moment **M** of gravity, i.e. it varies. As for the projection L_z, it remains constant since the vector **M** is perpendicular to the z axis and $M_z = 0$.

Now let us find analytical expressions for L_z and M_z. It is easy to see that this problem reduces to the determination of the projections of the vector products $[\mathbf{rp}]$ and $[\mathbf{rF}]$ on the z axis.

We shall make use of the cylindrical coordinate system ρ, φ, z, fixing the unit vectors $\mathbf{e}_\rho$, $\mathbf{e}_\varphi$, $\mathbf{e}_z$, oriented in the direction of increasing coordinates, to the particle A (Fig. 81). In this coordinate system the radius vector **r** and momentum **p** of the particle are written in the form

$$\mathbf{r} = \rho\mathbf{e}_\rho + z\mathbf{e}, \quad \mathbf{p} = p_\rho\mathbf{e}_\rho + p_\varphi\mathbf{e}_\varphi + p_z\mathbf{e}_z,$$

where p_ρ, p_φ, p_z are the projections of the vector **p** on the corresponding unit vectors. It is known from vector algebra

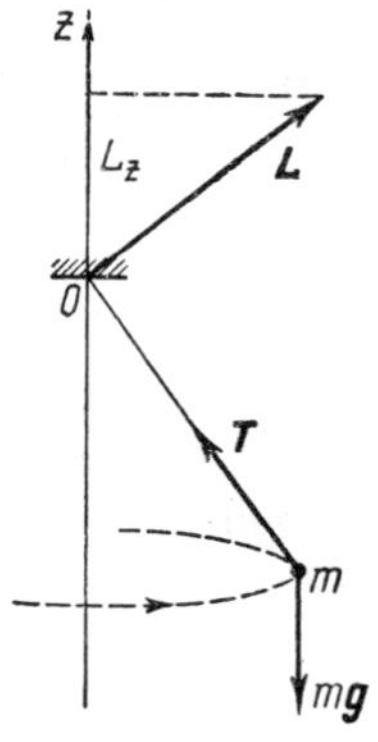

Fig. 80

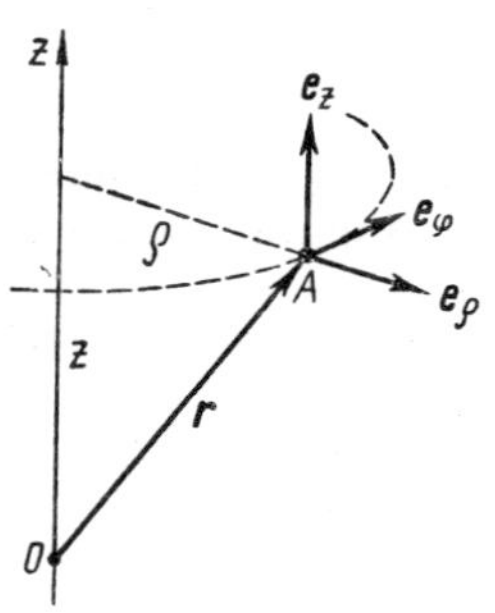

Fig. 81

that the vector product [**rp**] can be represented via the following determinant:

$$\mathbf{L} = [\mathbf{rp}] = \begin{vmatrix} \mathbf{e}_\rho & \mathbf{e}_\varphi & \mathbf{e}_z \\ \rho & 0 & z \\ p_\rho & p_\varphi & p_z \end{vmatrix}$$

From this it is immediately seen that the angular momentum of the particle relative to the z axis is

$$L_z = \rho p_\varphi, \tag{5.8}$$

where ρ is the distance of the particle from the z axis. Let us reduce this expression to a form more suitable for practical applications. Taking into account that $p_\varphi = mv_\varphi = m\rho\omega_z$, we get

$$L_z = m\rho^2\omega_z, \tag{5.9}$$

where ω_z is the projection of the angular velocity $\boldsymbol{\omega}$ with which the radius vector of the particle rotates.

The force moment relative to the z axis is written similarly to Eq. (5.8):

$$M_z = \rho F_\varphi, \tag{5.10}$$

where F_φ is the projection of the force vector $\mathbf{F}$ on the unit vector $\mathbf{e}_\varphi$.

Note that the projections L_z and M_z are indeed independent of the choice of the point O on the z axis, relative to which the vectors $\mathbf{L}$ and $\mathbf{M}$ are defined. Besides, it is seen that L_z and M_z are algebraic quantities, their signs corresponding to those of the projections p_φ and F_φ.

§ 5.2. The Law of Conservation of Angular Momentum

Let us choose an arbitrary system of particles and introduce the notion of the angular momentum of that system as the vector sum of angular momenta of its individual particles:

$$\mathbf{L} = \sum \mathbf{L}_i, \tag{5.11}$$

where all vectors are determined relative to the same point O of a given reference frame. Note that the angular momentum of the system is an *additive* quantity: the angular momentum of a system is equal to the sum of the angular momenta of its individual parts, irrespective of whether they interact or not.

Let us clarify what quantity defines the change of the angular momentum of the system. For this purpose we differentiate Eq. (5.11) with respect to time: $d\mathbf{L}/dt = \sum d\mathbf{L}_i/dt$. In the previous section it was shown that the derivative $d\mathbf{L}_i/dt$ is equal to the moment of all forces acting on the ith particle. We represent this moment as the sum of the moments of internal and external forces, i.e. $\mathbf{M}_i' + \mathbf{M}_i$. Then

$$d\mathbf{L}/dt = \sum \mathbf{M}_i' + \sum \mathbf{M}_i.$$

Here the first sum is the total moment of all internal forces relative to the point O and the second sum is the total moment of all external forces relative to the same point.

Let us demonstrate that the *total moment of all internal forces* relative to any point is *equal to zero*. Indeed, the internal forces are the forces of interaction between the particles of the given system. In accordance with Newton's third law these forces are pairwise equal in magnitude, are opposite in direction and lie on the same straight line, that is, have the same arm. Consequently, the force moments of each pair of interaction are equal in magnitude and opposite in direction, i.e. they counterbalance each other, and hence the total moment of all internal forces always equals zero.

As a result, the last equation takes the form

$$\boxed{d\mathbf{L}/dt=\mathbf{M},} \tag{5.12}$$

where $\mathbf{M}$ is the total moment of all *external* forces, $\mathbf{M} = \sum \mathbf{M}_i$.

Eq. (5.12) thus asserts that *the time derivative of the angular momentum of a system is equal to the total moment of all external forces*. It is understood that both quantities, $\mathbf{L}$ and $\mathbf{M}$, are determined relative to the same point O of a given reference frame.

As in the case of a single particle, from Eq. (5.12) it follows that the increment of the angular momentum of a system during the finite time interval t is

$$\boxed{\mathbf{L}_2-\mathbf{L}_1=\int_0^t \mathbf{M}\,dt,} \tag{5.13}$$

i.e. the increment of the angular momentum of a system is equal to the momentum of the total moment of all external forces during the corresponding time interval. Of course, the two quantities, $\mathbf{L}$ and $\mathbf{M}$, are also determined here relative to the same point O of a chosen reference frame.

Eqs. (5.12) and (5.13) are valid both in inertial and non-inertial reference frames. However, in a non-inertial reference frame one has to take into account the inertial forces acting as external forces, i.e. in these equations $\mathbf{M}$ should be regarded as the sum $\mathbf{M}_{ia}+\mathbf{M}_{in}$, where $\mathbf{M}_{ia}$ is the total moment of all external forces of interaction and $\mathbf{M}_{in}$ is the total moment of inertial forces (relative to the same point O of the reference frame).

Thus, we have reached the following significant conclusion: in accordance with Eq. (5.12) *the angular momentum of a system can change only due to the total moment of all external forces*. From this immediately follows another important conclusion, **the law of conservation of angular momentum:**

in an inertial reference frame the angular momentum of a closed system of particles remains constant, i.e. does not change with time. This statement is valid for an angular momentum determined relative to *any* point of the inertial reference frame.

Thus, in an inertial reference frame the angular momentum of a closed system of particles is

$$\boxed{\mathbf{L}=\sum \mathbf{L}_i(t)=\text{const.}} \qquad (5.14)$$

At the same time the angular momenta of individual parts or particles of a closed system can vary with time, a fact emphasized in the last expression. These variations, however, occur in such a way that the increment of the angular momentum in one part of the system is equal to the angular momentum decrease in another part (of course, relative to the same point of the reference frame).

In this respect Eqs. (5.12) and (5.13) can be regarded as a more general formulation of the angular momentum conservation law, a formulation specifying the *cause* of variation of the angular momentum of a system, which is the influence of other bodies (via the moment of external forces of interaction). All this, of course, is valid only in inertial reference frames.

Once again we shall point out the following: the law of conservation of angular momentum is valid only in inertial reference frames. This, however, does not rule out cases when the angular momentum of a system remains constant in non-inertial reference frames as well. For this, it is sufficient that, in accordance with Eq. (5.12), which holds true also in non-inertial reference frames, the total moment of all external forces (including inertial forces) be equal to zero. Such circumstances are realized very seldom, and the corresponding cases are exceptional.

The law of conservation of angular momentum is just as important as the energy and momentum conservation laws. In many cases this law by itself enables us to draw important inferences about the essential aspects of particular processes without going into their detailed analysis. We shall illustrate this by the following example.

Example. Two identical spheres are mounted on a smooth horizontal bar along which they can slide (Fig. 82). Initially the spheres are brought together and connected by a thread. Then the whole assembly

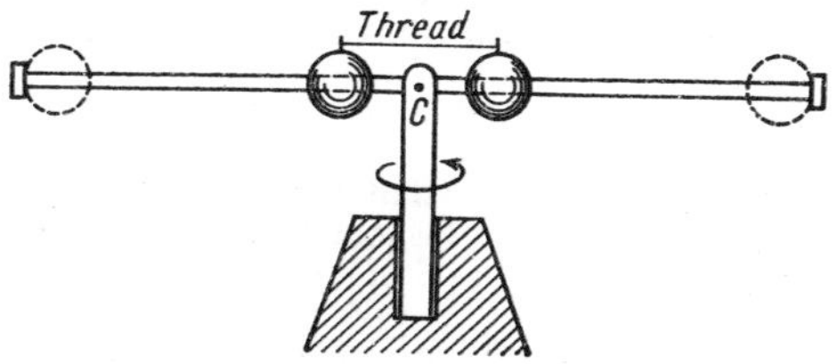

Fig. 82

is set into rotation about a vertical axis. After a period of free rotation, the thread is burned up. Naturally, the spheres fly apart toward the ends of the bar. At the same time, the angular velocity of the assembly drops drastically.

The observed phenomenon is a direct consequence of the law of conservation of angular momentum, for this assembly behaves as a closed system (the external forces counterbalance one another and the friction forces in the axis are small). To assess quantitatively the angular velocity change, let us assume the mass of the whole assembly to be concentrated in the spheres and their size to be negligible. Then, from the equality of the angular momenta of the spheres relative to the point C in the initial and final states of the system $2m\,[\mathbf{r}_1\mathbf{v}_1] = 2m\,[\mathbf{r}_2\mathbf{v}_2]$ it follows that

$$r_1^2\omega_1 = r_2^2\omega_2.$$

It is seen that as the distance r from the spheres to the rotation axis grows, the angular momentum of the assembly decreases (as $1/r^2$). And vice versa, if the distance between the spheres decreases (due to some internal forces), the angular velocity of the assembly increases. This general phenomenon is widely used by, for example, figure skaters and gymnasts.

Note that the final result is quite independent of the nature of internal forces (here the friction forces between the spheres and the bar).

Of special interest are cases in which the angular momentum **L** remains constant in non-closed systems in which the momentum **p** is known to change with time. If the total moment of external forces relative to a point O of the chosen reference frame $\mathbf{M} \equiv 0$ during the time interval considered, then in accordance with Eq. (5.12) the angular momentum of the system relative to the point O remains constant during this time interval. Generally speaking, such a point may not exist in non-closed systems, so that the fact of its existence should be established in every concrete case.

Example 1. The Earth-Moon system moving in the gravitational field of the Sun is non-closed. Its momentum continuously varies due to gravitational forces. However, there is one point here relative to which the moment of the gravitational forces acting on this system is always equal to zero. This point is the centre of the Sun. Therefore, it can be immediately claimed that the angular momentum of the Earth-Moon system relative to the Sun's centre remains constant.

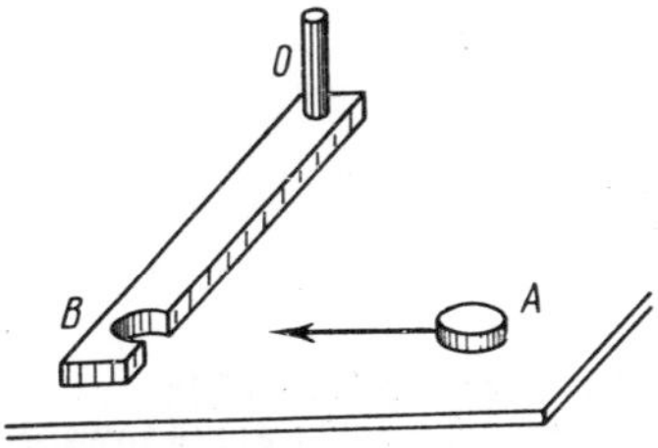

Fig. 83

Example 2. A rod OB lying on a smooth horizontal plane can rotate freely about a stationary vertical axis passing through the rod's end O (Fig. 83). A disc A moving along the plane hits the rod's end B and gets stuck there, whereupon the whole system starts rotating as a single unit about the point O.

It is clear that the disc and the rod compose a non-closed system: apart from the forces counterbalancing each other in the vertical direction, a horizontal force exerted by the axis is generated during the impact, while during the rotation the axis exerts a force impelling the centre of inertia of the system to move along the circle. But both of these forces pass through the point O, and therefore the moment of these external forces is always equal to zero (relative to the point O). Hence, the following conclusion can be drawn: the angular momentum of this system remains constant relative to the point O.

Infrequently in non-closed systems it is not the angular momentum **L** itself that remains constant, but its projection on a stationary axis z. This happens when the projection of the total moment **M** of all external forces on that axis z is equal to zero. In fact, projecting Eq. (5.12) on the z axis, we obtain

$$dL_z/dt = M_z. \qquad (5.15)$$

Here L_z and M_z are the angular momentum and the total moment of external forces relative to the z axis:

$$L_z = \sum L_{iz}, \quad M_z = \sum M_{iz}, \tag{5.16}$$

where L_{iz} and M_{iz} are the angular momentum and the moment of external forces relative to the z axis for the ith particle of the system.

It follows from Eq. (5.15) that if $\mathbf{M}_z \equiv 0$ relative to some stationary axis z in a given reference frame, the angular momentum of the system relative to that axis does not change:

$$L_z = \sum L_{iz}(t) = \text{const.} \tag{5.17}$$

At the same time the vector $\mathbf{L}$, defined relative to an arbitrary point O on that axis, may vary. For example, when a system moves in a uniform gravitational field, the total moment of all gravitational forces relative to any stationary point O is perpendicular to the vertical, and therefore $M_z \equiv 0$ and $L_z = \text{const}$ relative to any vertical axis. This cannot be said about the vector $\mathbf{L}$ itself.

The reasoning leading to the law of conservation of angular momentum is based entirely on the validity of Newton's laws. But what about systems that do not obey those laws, e.g. the systems with electromagnetic radiation, atoms, nuclei, etc.?

Because of the immense role that the law of conservation of angular momentum plays in mechanics, the concept of angular momentum is extended in physics to non-mechanical systems (which do not obey Newton's law) and the law of conservation of angular momentum is postulated for all physical processes.

The law of conservation of angular momentum thus extended is no longer a consequence of Newton's laws; it *represents an independent general principle generalized from experimental facts.* Together with the energy and momentum conservation laws *the law of conservation of angular momentum is one of the most important, fundamental laws of nature.*

§ 5.3. Internal Angular Momentum

It was established in the foregoing section that the angular momentum **L** of a system changes only due to the total moment **M** of all external forces; it is the vector **M** that defines the behaviour of the vector **L**. Now we shall examine the most essential properties of these quantities together with the significant conclusions following from those properties.

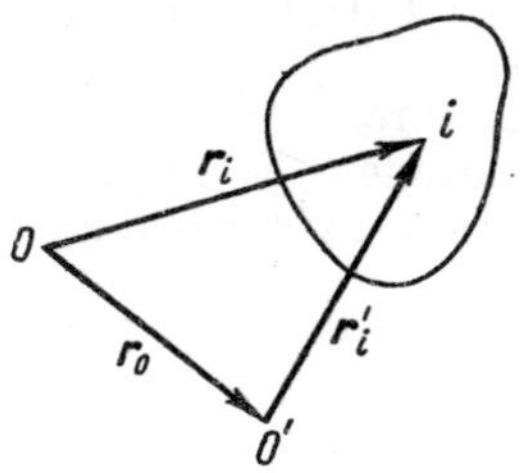

Fig. 84

The total moment of external forces. Just as the moment of an individual force, the total moment of forces depends, generally speaking, on the choice of a point relative to which the moment is determined. Let **M** be the total moment of forces relative to the point O and $\mathbf{M}'$ relative to the point O' whose radius vector is equal to $\mathbf{r}_0$ (Fig. 84). Let us find the relationship between **M** and $\mathbf{M}'$.

The radius vectors $\mathbf{r}_i$ and $\mathbf{r}'_i$ of the point at which the force $\mathbf{F}_i$ is applied are related as $\mathbf{r}_i = \mathbf{r}'_i + \mathbf{r}_0$ (Fig. 84). Consequently, **M** may be written in the following form:

$$\mathbf{M} = \sum [\mathbf{r}_i \mathbf{F}_i] = \sum [\mathbf{r}'_i \mathbf{F}_i] + \sum [\mathbf{r}_0 \mathbf{F}_i]$$

or

$$\mathbf{M} = \mathbf{M}' + [\mathbf{r}_0 \mathbf{F}], \tag{5.18}$$

where $\mathbf{F} = \sum \mathbf{F}_i$ is the resultant of all external forces.

Eq. (5.18) shows that when $\mathbf{F} = 0$, the total moment of external forces does not depend on the choice of the point relative to which it is determined. In particular, such is the case when a *couple* acts on a system.

In this respect the C frame possesses one interesting and important characteristic (recall that this reference frame is rigidly fixed to the centre of inertia of a system of particles and translates with respect to inertial frames). Since in the general case the C frame is non-inertial, the resultant of all external forces must include not only the external forces of interaction $\mathbf{F}_{ia}$ but also the inertial forces $\mathbf{F}_{in}$. On the other hand, the system of particles as a whole is at rest in the C frame, and therefore in accordance with Eq. (4.14) $\mathbf{F} = \mathbf{F}_{ia} + \mathbf{F}_{in} = 0$. Taking into account Eq. (5.18), we reach the following significant conclusion: *in the C frame the total moment of all external forces, including inertial forces, does not depend on the choice of the point O.*

And here is another important conclusion: *in the C frame the total moment of inertial forces relative to the centre of inertia is always equal to zero*:

$$\boxed{\mathbf{M}_C^{in} = 0.} \qquad (5.19)$$

Indeed, the inertial force acting on each particle of the system $\mathbf{F}_i = -m_i\mathbf{w}_0$, where $\mathbf{w}_0$ is the acceleration of the C frame. Consequently, the total moment of all these forces relative to the centre of inertia C is

$$\mathbf{M}_C^{in} = \sum [\mathbf{r}_i, \; -m_i\mathbf{w}_0] = -[(\sum m_i\mathbf{r}_i), \; \mathbf{w}_0].$$

In accordance with Eq. (4.8) $\sum m_i\mathbf{r}_i = m\mathbf{r}_C$, and as in our case $\mathbf{r}_C = 0$, then $\mathbf{M}_C^{in} = 0$.

Internal angular momentum. Generally speaking, angular momentum, just as the force moment, depends on the choice of the point O relative to which it is determined. When that point is transferred by the distance $\mathbf{r}_0$ (Fig. 84), the new radius vectors $\mathbf{r}'_i$ of the particles are related to the old ones $\mathbf{r}_i$ by means of the formula $\mathbf{r}_i = \mathbf{r}'_i + \mathbf{r}_0$. Consequently, the angular momentum of the system relative to the point O can be written as follows:

$$\mathbf{L} = \sum [\mathbf{r}_i\mathbf{p}_i] = \sum [\mathbf{r}'_i\mathbf{p}_i] + \sum [\mathbf{r}_0\mathbf{p}_i],$$

or

$$\mathbf{L} = \mathbf{L}' + [\mathbf{r}_0\mathbf{p}], \qquad (5.20)$$

where $\mathbf{L}'$ is the angular momentum of the system relative to the point O', and $\mathbf{p} = \sum \mathbf{p}_i$ is the total momentum of the system.

From Eq. (5.20) it follows that if the total momentum of the system $\mathbf{p} = 0$, then its angular momentum does not depend on the choice of the point O. This distinguishes the C frame, in which the system of particles as a whole is at rest. Hence, we reach the third important conclusion: *in the C frame the angular momentum of a system of particles is independent of the choice of the point* relative to which it is determined. We shall refer to this quantity as the *internal angular momentum* of a system and denote it by $\tilde{\mathbf{L}}$.

Relation between L and $\tilde{\mathbf{L}}$. Suppose $\mathbf{L}$ is the angular momentum of a system of particles relative to the point O of the K reference frame. Since the internal angular momentum $\tilde{\mathbf{L}}$ in the C frame does not depend on the choice of the point O', this point may be taken coincident with the point O of the K frame *at a given moment* of time. Then at that moment the radius vectors of all the particles in both reference frames are equal ($\mathbf{r}_i' = \mathbf{r}_i$) and the velocities are related by the formula

$$\mathbf{v}_i = \tilde{\mathbf{v}}_i + \mathbf{V}_C, \qquad (5.21)$$

where $\mathbf{V}_C$ is the velocity of the C frame relative to the K frame. Consequently, we may write

$$\mathbf{L} = \sum m_i [\mathbf{r}_i \mathbf{v}_i] = \sum m_i [\mathbf{r}_i \tilde{\mathbf{v}}_i] + \sum m_i [\mathbf{r}_i \mathbf{V}_C]. \qquad (5.22)$$

The first sum on the right-hand side of this equality is the internal angular momentum $\tilde{\mathbf{L}}$. The second sum may be written in accordance with Eq. (4.8) as $m\,[\mathbf{r}_C \mathbf{V}_C]$, or $[\mathbf{r}_C \mathbf{p}]$, where m is the mass of the whole system, $\mathbf{r}_C$ is the radius vector of its centre of inertia in the K frame and $\mathbf{p}$ is the total momentum of the system. Finally, we obtain

$$\boxed{\mathbf{L} = \tilde{\mathbf{L}} + [\mathbf{r}_C \mathbf{p}],} \qquad (5.23)$$

i.e. *the angular momentum* $\mathbf{L}$ *of a system of particles comprises its internal angular momentum* $\tilde{\mathbf{L}}$ *and the momentum* $[\mathbf{r}_C \mathbf{p}]$, *associated with the motion of the system of particles as a single unit.*

Let us consider, for example, a uniform sphere rolling down an inclined plane. Its angular momentum relative to some point of that plane is composed of the angular momentum associated with the motion of the centre of inertia of the sphere and the internal angular momentum associated with the rotation of the sphere about its axis.

Specifically, it follows from Eq. (5.23) that if the centre of inertia of a system is at rest (the momentum of the system $\mathbf{p} = 0$), then its angular momentum $\mathbf{L}$ represents the internal angular momentum. We are already familiar with that case. In another extreme case, when $\widetilde{\mathbf{L}} = 0$, the angular momentum of a system relative to some point is determined only via the momentum associated with the motion of the system as a single unit, i.e. by the second term of Eq. (5.23). This is how, for example, the angular momentum of any solid behaves during its translation.

Equation of moments in the C frame. We pointed out in the previous section that Eq. (5.12) holds true in any reference frame. Consequently, it is valid in the C frame as well, and we can immediately write: $d\widetilde{\mathbf{L}}/dt = \widetilde{\mathbf{M}}$, where $\widetilde{\mathbf{M}}$ is the total moment of external forces in the C frame.

Since the C frame is non-inertial in the general case, $\widetilde{\mathbf{M}}$ includes not only the moments of external forces of interaction, but also the moment of inertial forces. On the other hand, at the beginning of this section (see p. 161) the force moment $\widetilde{\mathbf{M}}$ in the C frame was shown to be independent of the choice of the point relative to which it is determined. Usually, the point C, the centre of inertia of the system, is taken as such a point. The choice of this point is advantageous because the total moment of inertial forces relative to it is equal to zero, so that one must take into account *only* the total moment $\mathbf{M}_C$ of external forces of interaction. Thus,

$$\boxed{d\widetilde{\mathbf{L}}/dt = \mathbf{M}_C,} \tag{5.24}$$

i.e. the time derivative of the internal angular momentum of a system is equal to the total moment of all external forces of interaction relative to the centre of inertia of that system.

In particular, when $\mathbf{M}_C \equiv 0$, then $\widetilde{\mathbf{L}} = \text{const}$, i.e. the *internal angular momentum* of a system does not vary.

When written in projections on the z axis passing through the centre of inertia of a system, Eq. (5.24) takes the form

$$d\widetilde{L}_z/dt = M_{Cz} \tag{5.25}$$

where M_{Cz} is the total moment of external forces of interaction relative to the z axis fixed in the C frame and passing through the centre of inertia. Here again, if $M_{Cz} \equiv 0$, then $\widetilde{L}_z = \text{const}$.

§ 5.4. Dynamics of a Solid

In the general case the motion of a solid is defined by two vector equations. One of them is the equation of motion of the centre of inertia (4.11), and the other the equation relating momenta and moments in the C frame, Eq. (5.24):

$$m\, d\mathbf{V}_C/dt = \mathbf{F}; \quad d\tilde{\mathbf{L}}/dt = \mathbf{M}_C. \tag{5.26}$$

If the laws of acting external forces, the points of their application and the initial conditions are known, these equations provide the velocity and the position of each point of a solid at any moment of time, i.e. make it possible to solve the problem of motion of a solid completely. However, despite the apparent simple form of Eqs. (5.26), their solution in the general case is far from easy. First of all this is because the relationship between the internal angular momentum $\tilde{\mathbf{L}}$ and the velocities of individual points of a solid in the C frame turns out to be complicated, except for a few special cases. We shall not consider this problem in the general case (it is solved in the general theory) and shall limit ourselves hereafter to only several special cases.

But first we shall quote some considerations which follow directly from the very appearance of Eqs. (5.26). Clearly, translation of forces along the direction of their action does not affect either the resultant $\mathbf{F}$ or the total moment $\mathbf{M}_C$. Eqs. (5.26) do not vary either, and therefore the motion of a solid remains the same. Consequently, the points of application of external forces can be transferred along the direction of their action, a technique used very extensively.

Equivalent force. In those cases when the total moment of all external forces turns out to be perpendicular to the resultant force, i.e. $\mathbf{M} \perp \mathbf{F}$, all external forces may be reduced to *one* force $\mathbf{F}$ acting along a certain straight line. In fact, if the total moment relative to some point O $\mathbf{M} \perp \mathbf{F}$, then we can always find a vector $\mathbf{r}_0 \perp \mathbf{M}$ (Fig. 85), such that with the given $\mathbf{M}$ and $\mathbf{F}$

$$\mathbf{M} = [\mathbf{r}_0 \mathbf{F}].$$

Here the choice of $\mathbf{r}_0$ is not unambiguous: adding any vector $\mathbf{r}$ parallel to $\mathbf{F}$, we do not violate the last equality. This means that this equality defines not a point of "application" of the force **F**, but a line along which it acts. Knowing the moduli M and F of the corresponding vectors, we can find the arm l of the force **F** (Fig. 85): $l = M/F$.

Thus, if $\mathbf{M} \perp \mathbf{F}$, then the system of forces acting on various points of a solid may be replaced by one *equivalent* force which is equal to the resultant force **F** and produces a moment equal to the total moment **M** of all external forces.

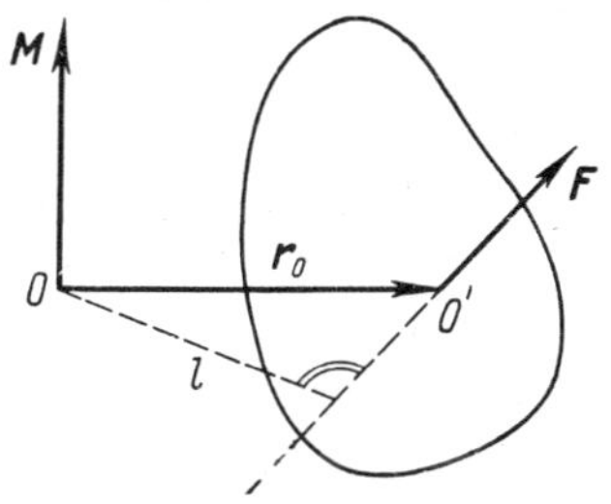

Fig. 85

A uniform field of force, e.g. the gravitational field, can serve as an example. In such a field each particle experiences the force $\mathbf{F}_i = m_i\mathbf{g}$. In this case the total moment of gravitational forces relative to any point O is equal to

$$\mathbf{M} = \sum [\mathbf{r}_i, m_i\mathbf{g}] = [(\sum m_i\mathbf{r}_i)\,\mathbf{g}].$$

In accordance with Eq. (4.8) the sum in parentheses is equal to $m\mathbf{r}_C$, where m is the mass of a body and $\mathbf{r}_C$ is the radius vector of its centre of inertia relative to the point O. Therefore,

$$\mathbf{M} = [m\mathbf{r}_C,\ \mathbf{g}] = [\mathbf{r}_C,\ m\mathbf{g}].$$

This implies that the equivalent force $m\mathbf{g}$ of the gravitational forces passes through the centre of inertia of the body. It is customary to say that the equivalent force of gravity is "applied" to the centre of inertia of a body, or to its centre of gravity. Clearly, the moment of this force relative to the centre of inertia of a body is equal to zero.

Now we shall move on to an examination of the four specific cases of motion of a solid: (1) rotation about a stationary axis, (2) plane (two-dimensional) motion, (3) rotation about free axes, (4) the special case of motion when a body has only one motionless point (gyroscopes).

1. Rotation about a stationary axis. First let us find the expression for the angular momentum of a solid body relative to the rotation axis OO' (Fig. 86). Making use of Eq. (5.9), we write

$$L_z = \sum L_{iz} = (\sum m_i\rho_i^2)\,\omega_z,$$

where m_i and ρ_i are the mass of the ith particle of a solid and its distance from the rotation axis, and ω_z is its angular velocity. Denoting the quantity in parentheses by I, we get

$$\boxed{L_z = I\omega_z,} \tag{5.27}$$

where I is the so-called *moment of inertia* of a solid relative to the OO' axis:

$$I = \sum m_i \rho_i^2. \tag{5.28}$$

It is easy to see that the moment of inertia of a solid depends on the distribution of masses relative to the axis in question, and is an additive quantity.

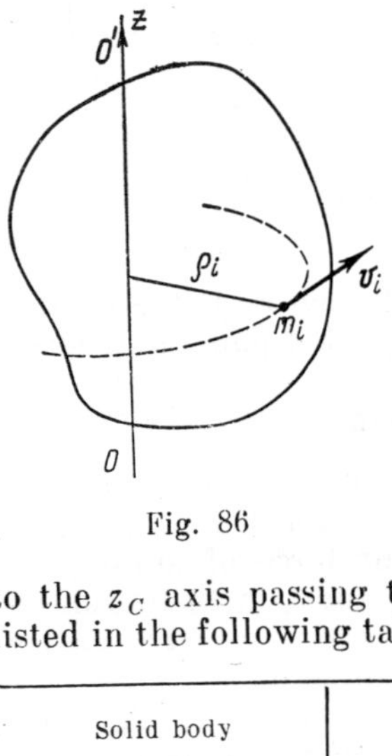

Fig. 86

The moment of inertia of a body is calculated by means of the following formula:

$$I = \int r^2\, dm = \int \rho r^2\, dV,$$

where dm and dV are the mass and the volume of an element of the body located at the distance r from the z axis chosen, and ρ is the density of the body at a given point.

The moments of inertia of some uniform solids relative to the z_C axis passing through their centres of inertia are listed in the following table (here m is the mass of the body):

Solid body	z_C axis	Moment of inertia I_C
A thin rod of length l	Is perpendicular to the rod	$\frac{1}{12}ml^2$
A uniform cylinder of radius R	Coincides with the axis of the cylinder	$\frac{1}{2}mR^2$
A thin disc of radius R	Coincides with the diameter of the disc	$\frac{1}{4}mR^2$
A sphere of radius R	Passes through the centre of the sphere	$\frac{2}{5}mR^2$

In terms of mathematics the calculation of the moment of inertia of a solid body of an arbitrary shape relative to one or another axis presents, generally speaking, a painstaking task. In certain cases, however, the moment of inertia can be calculated much more easily if one resorts to the *Steiner theorem*: the moment of inertia I relative to an arbitrary z axis is equal to the moment of inertia I_C relative to the z_C axis which is parallel to the z axis and passes through the centre of inertia C of the body, plus the product of the mass m of the body by the square of the distance a between the axes:

$$\boxed{I = I_C + ma^2.} \tag{5.29}$$

The theorem is proved in Appendix 3.

Thus, when the moment of inertia I_C is known, the moment of inertia I is calculated with no effort. For example, the moment of inertia of a thin rod (of mass m and length l) relative to an axis perpendicular to the rod and passing through its end is equal to

$$I = \frac{1}{12} ml^2 + m\left(\frac{l}{2}\right)^2 = \frac{1}{3} ml^2.$$

The fundamental equation of rotation dynamics of a solid body (stationary axis of rotation). This equation can be easily obtained as a consequence of Eq. (5.15) by differentiating Eq. (5.27) with respect to time. Then

$$\boxed{I\beta_z = M_z,} \tag{5.30}$$

where M_z is the total moment of all external forces relative to the rotation axis. Specifically, from this equation the moment of inertia I is seen to determine the inertial properties of a solid body during its rotation: the same moment of forces M_z induces a smaller angular acceleration β_z in bodies possessing greater moments of inertia.

Recall that force moments relative to an axis are algebraic quantities: their signs depend on both the choice of the positive direction of the z axis (coinciding with the rotation axis) and the "rotation" direction of the corresponding force moment. For example, choosing the positive direction of the z axis as shown in Fig. 87, we thereby specify the positive direction of reading the angle φ (both of these directions are associated with the right-hand screw rule). Next, if a certain moment M_{iz} "rotates" a body in the positive direction of the angle φ, that moment is regarded as positive, and vice versa. In its turn, the sign of the total moment M_z determines the sign of β_z, the projection of the angular acceleration vector on the z axis.

By integrating Eq. (5.30) with allowance made for the initial conditions, the values ω_{0z} and φ_0 at the initial moment of time, we can obtain a comprehensive solution of the problem of a solid body rotating about a stationary axis, i.e. obtain the time dependence of the angular velocity $\omega_z(t)$ and the rotation angle $\varphi(t)$.

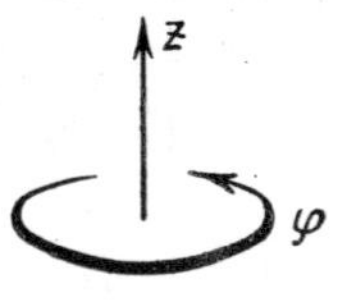

Fig. 87

Note that Eq. (5.30) is valid in *any* reference frame fixed rigidly to the rotation axis. However, if the reference frame is non-inertial, it should be borne in mind that the force moment M_z consists not only of moments of forces of interaction with other bodies, but also moments of inertial forces.

Kinetic energy of a rotating solid body (stationary axis of rotation). Since the velocity of the ith particle of a rotating solid body is $v_i = \rho_i\omega$, we may write

$$T = \sum m_i v_i^2/2 = \left(\sum m_i\rho_i^2\right)\omega^2/2$$

or briefly,

$$\boxed{T = I\omega^2/2,} \tag{5.31}$$

where I is the moment of inertia of the body relative to the rotation axis and ω is its angular velocity.

Example. Disc *1* (Fig. 88) rotates about a smooth vertical axis with the angular velocity $\boldsymbol{\omega}_1$. Disc *2* rotating with the angular velocity $\boldsymbol{\omega}_2$ falls on disc *1*. Due to friction between them the discs soon start rotating as a single unit. Find the increment of the rotational kinetic energy of that system provided that the moments of inertia of the discs relative to the rotation axis are equal to I_1 and I_2 respectively.

First let us find the steady-state angular velocity of rotation. From the law of conservation of the angular momentum of a system relative to the z axis it follows that $I_1\omega_{1z} + I_2\omega_{2z} = (I_1 + I_2)\,\omega_z$, whence

$$\omega_z = (I_1\omega_{1z} + I_2\omega_{2z})/(I_1 + I_2). \quad (1)$$

Note that ω_{1z}, ω_{2z} and ω_z are algebraic quantities. If $\omega_z > 0$, the corresponding vector ω coincides with the positive direction of the z axis, and vice versa.

The increment of the kinetic energy of rotation for this system is

$$\Delta T = (I_1 + I_2)\,\omega_z^2/2 - (I_1\omega_{1z}^2/2 + I_2\omega_{2z}^2/2).$$

Replacing ω_z by its expression (1), we get

$$\Delta T = -\frac{I_1 I_2}{2(I_1 + I_2)}(\omega_{1z} - \omega_{2z})^2. \quad (2)$$

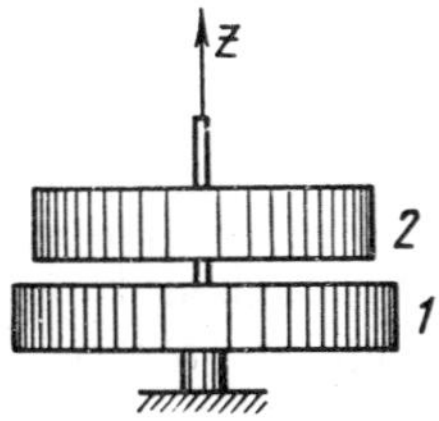

Fig. 88

The minus sign shows that the kinetic energy of the system decreases.

Note that the results (1) and (2) are quite similar in form and meaning to the case of absolutely inelastic collision (see p. 127).

The work performed by external forces during the rotation of a solid body about a stationary axis. In accordance with the law of variation of mechanical energy of a system the elementary work of all external forces acting on a solid body is equal only to the increment of kinetic energy of the body, as its internal potential energy remains constant, i.e. $\delta A = dT$. Making use of Eq. (5.31), we may write $\delta A = d\,(I\omega^2/2)$. Since the z axis coincides with the rotation axis, $\omega^2 = \omega_z^2$ and

$$\delta A = I\omega_z\,d\omega_z.$$

But in accordance with Eq. (5.30), $I d\omega_z = M_z\,dt$. Substituting this expression into the last equation for δA and taking into account that $\omega_z\,dt = d\varphi$, we obtain

$$\delta A = M_z\,d\varphi. \quad (5.32)$$

The work δA is an algebraic quantity: if M_z and $d\varphi$ have identical signs, then $\delta A > 0$; if the signs are opposite, then $\delta A < 0$.

The work performed by external forces during the rotation of a solid through an angle φ is equal to

$$\boxed{A = \int_0^{\varphi} M_z\,d\varphi.} \quad (5.33)$$

When M_z = const, the last expression takes a simple form: $A = M_z\varphi$.

Thus, the work performed by external forces during the rotation of a solid around a stationary axis is determined by the action of the moment M_z of these forces relative to that axis. If the forces are such that their moment $M_z \equiv 0$, then they perform no work.

2. Plane motion of a solid body (see p. 26). During plane motion the centre of inertia C of a solid body moves in a certain plane stationary in a given reference frame K while the angular velocity vector $\boldsymbol{\omega}$ of the body remains permanently perpendicular to that plane. This signifies that the body in the C frame performs a purely rotational motion about the stationary (in that frame) axis passing through the centre of inertia of the body. At the same time, the rotation of a solid is defined by Eq. (5.30), which was shown to be valid in any reference frame.

Thus, we have the following two equations describing plane motion of a solid body:

$$m\mathbf{w}_C = \mathbf{F}; \quad I_C\beta_z = M_{Cz}, \tag{5.34}$$

where m is the mass of the body, $\mathbf{F}$ is the resultant of all external forces, I_C and M_{Cz} are the moment of inertia and the total moment of all external forces, both moments being determined relative to the axis passing through the centre of inertia of the body.

It should be borne in mind that the moment M_{Cz} includes only external forces of interaction in spite of the fact that in the general case the C frame is non-inertial. This is because the total moment of inertial forces is equal to zero both relative to the centre of inertia and relative to any axis passing through that point. Therefore it can be disregarded altogether (see p. 163).

Note also that the angular acceleration β_z, as well as ω_z and φ, are equal in both reference frames since the C frame *translates* relative to the inertial reference frame K.

Integrating Eqs. (5.34) with allowance made for the initial conditions, we can find the relationships $\mathbf{r}_C(t)$ and $\varphi(t)$ defining the position of a solid body at any moment t.

When finding the motion of a *non-free* solid body one has

to use still another, additional condition specifying the restrictions that existing bonds impose on the motion. This condition provides a kinematic relationship between the linear and angular accelerations.

Example. A uniform cylinder of mass m and radius r rolls without slipping down an inclined plane forming the angle α with the horizontal (Fig. 89). Find the motion equation of the cylinder.

The *standard approach* to the solution of problems of this type consists in the following. First of all, one should identify the forces acting on a given body and the points of their application (in this case the acting forces include mg, gravity, $\mathbf{R}$, the normal component of the force of reaction of the inclined plane and $\mathbf{F}_{fr}$, the static friction force). Then it is necessary to choose the positive directions of the x axis and of the rotation angle φ (these directions should be consistent so that the accelerations w_{Cx} and β_z have the identical signs), e.g. as shown in Fig. 89, right. And only after that can the equations of motion (5.34) be written in terms of projections on the chosen positive directions of x and φ:

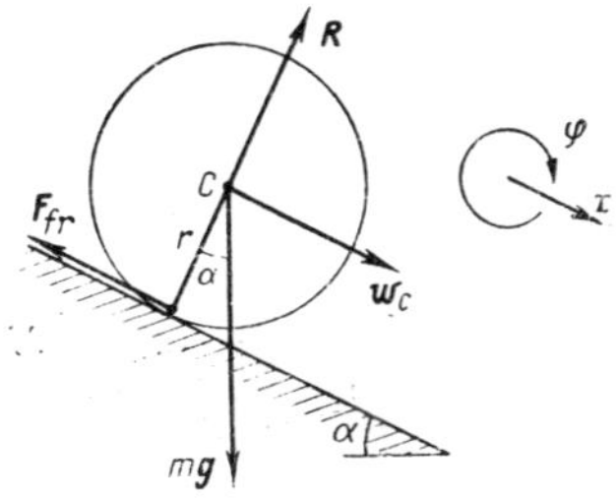

Fig. 89

$$mw_{Cx} = mg\sin\alpha - F_{fr}, \qquad I_C\beta_z = rF_{fr}.$$

In addition, the absence of slipping provides the kinematic relationship between the accelerations:

$$w_{Cx} = r\beta_z.$$

The simultaneous solution of all three equations allows the accelerations $\mathbf{w}_C$ and $\boldsymbol{\beta}$, as well as the $\mathbf{F}_{fr}$ force, to be found.

Kinetic energy in the plane motion of a solid body. Suppose a body performs a plane motion in a certain inertial reference frame K. To find its kinetic energy T in this frame, we shall resort to Eq. (4.12). The quantity $\tilde{T}$ entering into this equation represents in this case the kinetic energy of rotation of the body in the C frame about an axis passing through the body's centre of inertia. In accordance with Eq. (5.31) $\tilde{T} = I_C\omega^2/2$, therefore we may immediately write

$$\boxed{T = \frac{I_C \omega^2}{2} + \frac{mV_C^2}{2},} \tag{5.35}$$

where I_C is the moment of inertia of the body relative to the rotation axis passing through its centre of inertia, ω is the angular velocity of the body, m is its mass, and V_C is the velocity of the body's centre of inertia in the K reference frame.

Thus, *the kinetic energy of a solid body in a plane motion comprises the rotation energy in the C frame and the energy associated with the motion of the centre of inertia.*

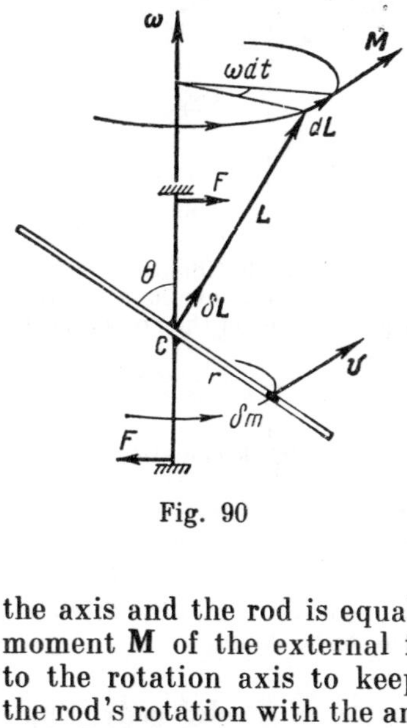

Fig. 90

3. Free axes. Principal axes of inertia. When a solid body is set in rotation and then left free, the direction of the rotation axis changes in space, generally speaking. To make the body's arbitrary rotation axis keep its direction constant, some forces need to be applied to that axis.

Let us consider this problem in detail using the following example. Suppose the middle point C of a uniform rod is rigidly fixed to a rotation axis so that the angle between the axis and the rod is equal to θ (Fig. 90). Let us find the moment $\mathbf{M}$ of the external forces which should be applied to the rotation axis to keep its direction constant during the rod's rotation with the angular velocity $\boldsymbol{\omega}$. In accordance with Eq. (5.12) this moment $\mathbf{M} = d\mathbf{L}/dt$. Thus, to determine $\mathbf{M}$, one should first find the angular momentum $\mathbf{L}$ of the rod and then its time derivative.

The angular momentum $\mathbf{L}$ can be determined most easily relative to the point C. Let us isolate mentally the rod's element of mass δm located at the distance $\mathbf{r}$ from the point C. Its angular momentum relative to that point $\delta\mathbf{L} = [\mathbf{r}, \delta m\mathbf{v}]$, where $\mathbf{v}$ is the velocity of the element. It can be easily seen from Fig. 90 that the vector $\delta\mathbf{L}$ is perpendicular to the rod and its direction is independent of the choice of the element δm. Consequently, the total angular momentum $\mathbf{L}$ of the rod coincides with the vector $\delta\mathbf{L}$ in direction.

Note that in this case the vector $\mathbf{L}$ does not coincide in its direction with the vector $\boldsymbol{\omega}$!

During the rotation of the rod the vector $\mathbf{L}$ also rotates with the angular velocity $\boldsymbol{\omega}$. In the time interval dt the vector $\mathbf{L}$ acquires the increment $d\mathbf{L}$, whose magnitude is seen from Fig. 90 to be equal to

$$|d\mathbf{L}| = L \sin(\pi/2 - \theta)\, \omega\, dt,$$

or, in a vector form, $d\mathbf{L} = [\boldsymbol{\omega}\mathbf{L}]\, dt$. Dividing both sides of the last expression by dt, we obtain

$$\mathbf{M} = [\boldsymbol{\omega}\mathbf{L}].$$

Thus, to maintain the rotation axis in a fixed direction, it should be subjected to the moment $\mathbf{M}$ of some external forces F (shown in Fig. 90). However, it is easy to see that when $\theta = \pi/2$, the vector $\mathbf{L}$ coincides in its direction with the vector $\boldsymbol{\omega}$, and in this case $\mathbf{M} \equiv 0$, i.e. the direction of the rotation axis remains invariable in the absence of external influence.

A rotation axis whose direction in space remains invariable in the absence of any external forces is referred to as a *free axis* of a body.

It is proved in the general theory that for any solid body there are three mutually perpendicular axes which pass through the centre of inertia of a body and can serve as free axes. They are called the *principal axes of inertia* of a body.

The determination of the principal axes of inertia of a body of an arbitrary form is a complex mathematical problem. It is much simpler, however, for bodies possessing some kind of symmetry, since the position of the centre of inertia and the direction of the principal axes of inertia possess in this case the same kind of symmetry.

For example, a uniform rectangular parallelepiped has principal axes passing through the centres of opposite faces. When a body possesses a symmetry axis (e.g. a uniform cylinder), one of its principal axes of inertia may be represented by that symmetry axis while any two mutually perpendicular axes lying in the plane perpendicular to the symmetry axis and passing through the body's centre of inertia can be chosen as the other axes. Thus, in a body possessing axial symmetry only one of the principal axes of inertia is fixed. In a body possessing central symmetry (e.g. in a uniform sphere) any three mutually perpendicular axes passing through the body's centre can be chosen as the principal axes of inertia, that is, not a single principal axis of inertia is fixed with respect to the body.

The important characteristic property of the principal axes of inertia of a body is the fact that during the rotation of the body about any of them the angular momentum **L** of the body coincides in its direction with the angular velocity $\boldsymbol{\omega}$ of the body and is determined by the formula

$$\mathbf{L} = I\boldsymbol{\omega}, \tag{5.36}$$

where I is the moment of inertia of the body relative to the given principal axis of inertia. Here **L** does not depend on the choice of the point relative to which it is determined (assuming the rotation axis to be stationary).

The validity of Eq. (5.36) may be easily demonstrated for the case of a uniform body possessing axial symmetry. Indeed, in accordance with Eq. (5.27) the angular momentum of a solid body relative to the rotation axis $L_z = I\omega_z$ (recall that L_z is the projection of the vector **L** determined relative to any point on that axis). But when a body is symmetric relative to the rotation axis, it immediately follows from the condition of symmetry that the vector **L** coincides in its direction with the vector $\boldsymbol{\omega}$, and therefore, $\mathbf{L} = I\boldsymbol{\omega}$.

Once again, it should be pointed out that in the general case (when the rotation axis does not coincide with any of the principal axes of inertia though it passes through the centre of inertia of a body) the direction of the vector **L** does not coincide with that of the vector $\boldsymbol{\omega}$, and the relationship between these vectors is very complex. This fact accounts for the complicated behaviour of rotating solids.

4. Gyroscopes. A gyroscope is a massive symmetric body rotating about its symmetry axis with a high angular velocity. Let us examine the behaviour of a gyroscope using a top as an example. It is known from experience that if the axis of a spinning top is tilted from the vertical, the top does not fall, but performs so-called *precession*, with its axis circumscribing a cone about the vertical with a certain angular velocity $\boldsymbol{\omega}'$. It turns out that with an increase in the angular velocity $\boldsymbol{\omega}$ of spinning of the top, the angular velocity $\boldsymbol{\omega}'$ of precession decreases.

Such behaviour of a gyroscope can be readily explained using the equation of momenta (5.12), assuming $\omega \gg \omega'$. This condition, by the way, elucidates what is meant by the *rapid* spinning of a gyroscope. In fact, the angular momentum **L** of a precessing gyroscope relative to the supporting point O (Fig. 91) may be represented as the sum of the angular momentum $\mathbf{L}_\omega$ associated with the gyroscope spinning about its axis and some additional angular momentum $\mathbf{L}'$ caused by the gyroscope precession about the vertical axis, i.e.

$$\mathbf{L} = \mathbf{L}_\omega + \mathbf{L}'.$$

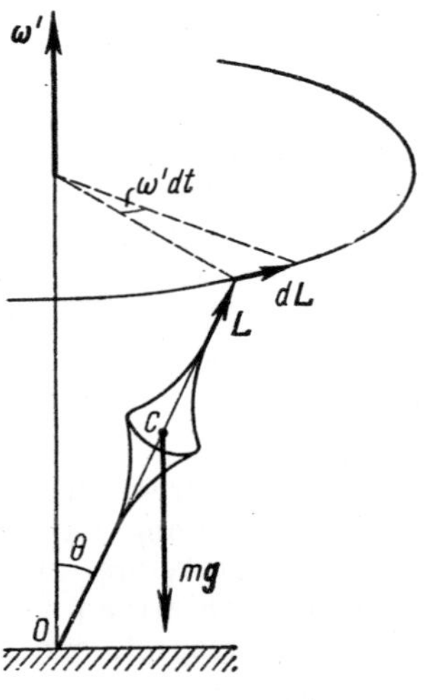

Fig. 91

Since the gyroscope axis coincides with one of the principal axes of inertia, then in accordance with Eq. (5.36) $\mathbf{L}_\omega = I\boldsymbol{\omega}$, where I is the moment of inertia of a gyroscope with respect to that axis. Moreover, it is clear that as the precession gets slower, the corresponding angular momentum $\mathbf{L}'$ diminishes. If $\omega \gg \omega'$, then in all practical cases $L_\omega \gg L'$, and therefore the resultant angular momentum **L** essentially coincides with $\mathbf{L}_\omega$ both in magnitude and direction. We can thus assume that

$$\mathbf{L} = I\boldsymbol{\omega}.$$

Knowing the behaviour of the vector **L**, we can find the motion characteristics of a gyroscope axis.

But the behaviour of the vector **L** is described by the equation of moments (5.12). In accordance with that equation the angular momentum **L** relative to the point O (Fig. 91) acquires during the time interval dt the increment

$$d\mathbf{L} = \mathbf{M}\, dt, \tag{5.37}$$

coinciding in direction with the vector **M**, the moment of external forces relative to the same point O (in this case that is the moment of the gravitational force $m\mathbf{g}$). From Fig. 91 it is seen that $d\mathbf{L} \perp \mathbf{L}$. As a result, the vector **L** (and, consequently, the gyroscope axis) spins together with the vector **M** about the vertical, circumscribing a circular cone with the half-aperture angle θ. The gyroscope precesses about the vertical axis with some angular velocity $\boldsymbol{\omega}'$.

Let us find how the vectors **M**, **L** and $\boldsymbol{\omega}'$ are interrelated. From the figure the modulus increment of the vector **L** during the time interval dt is seen to be equal to $|d\mathbf{L}| = L \sin\theta\, \omega' dt$, or in vector form $d\mathbf{L} = [\boldsymbol{\omega}'\mathbf{L}]\, dt$. Substituting this expression into Eq. (5.37), we obtain

$$[\boldsymbol{\omega}'\mathbf{L}] = \mathbf{M}. \tag{5.38}$$

It is seen from this equation that the moment of force **M** defines the angular precession *velocity* $\boldsymbol{\omega}'$ (but not acceleration!). Therefore, an instantaneous elimination of the moment **M** entails an instantaneous disappearance of precession. In this respect, one may say that precession possesses no inertia.

Note that the force moment **M** acting on a gyroscope may be very different in nature. Continuous precession, i.e. the constant angular velocity $\boldsymbol{\omega}'$, is maintained provided the vector **M** remains constant in magnitude and spins together with the gyroscope axis.

Example. Find the angular velocity of precession of a tilted gyroscope of mass m spinning with a high angular velocity $\boldsymbol{\omega}$ about its symmetry axis with respect to which the moment of inertia of the gyroscope is equal to I. The gyroscope's centre of inertia is located at a distance l from the supporting point.

In accordance with Eq. (5.38) $\omega' I\omega \sin\theta = mgl \sin\theta$, where θ is the angle between the vertical and the gyroscope axis (Fig. 91). Hence,

$$\omega' = mgl/I\omega.$$

It is interesting to note that the quantity ω' is independent of the inclination angle θ of the gyroscope axis. In addition, the result obtained shows that ω' is inversely proportional to ω, i.e. the higher the angular velocity of a gyroscope, the lower the angular velocity of its precession.

Gyroscopic couple. Now let us examine a phenomenon appearing with the forced rotation of a gyroscope axis. Suppose, for example, a gyroscope axis is mounted in a U-shaped support, which we rotate about the OO' axis as

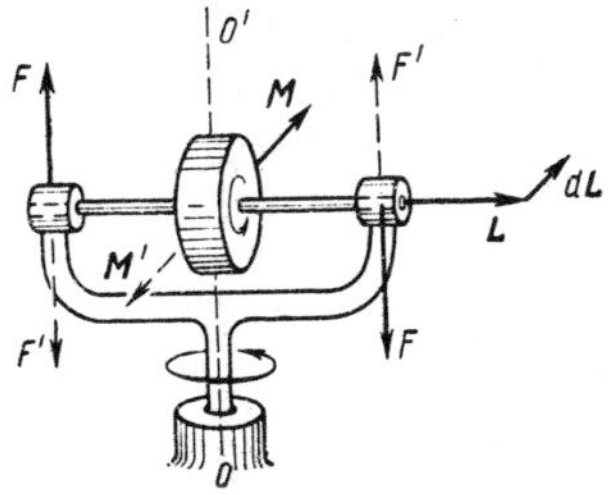

Fig. 92

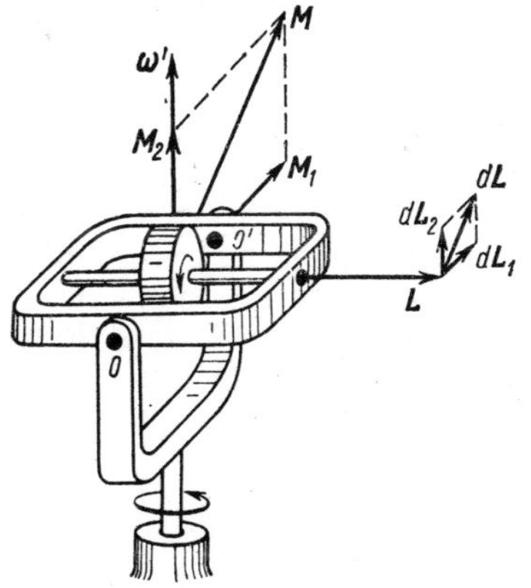

Fig. 93

shown in Fig. 92. If the angular momentum **L** of the gyroscope is directed to the right, then during the time interval dt in the process of that rotation the vector **L** gets an increment d**L**, a vector directed beyond the plane of the figure. In accordance with Eq. (5.37) this implies that the gyroscope experiences the force moment **M** coinciding with the d**L** vector in direction. The moment **M** is due to the appearance of the couple F that the support exerts on the gyroscope axis. In accordance with Newton's third law, on the other hand, the support itself experiences the forces F' developed by the gyroscope axis (Fig. 92). These forces are referred to as *gyroscopic*; they form the *gyroscopic couple* $\mathbf{M}' = -\mathbf{M}$. Note that in this case the gyroscope is not capable of resisting the variation of its rotation axis direction.

The appearance of gyroscopic forces is referred to as *gyrostatic action*, or gyroeffect. A gyroeffect associated with the appearance of gyroscopic stresses in bearings is observed, for example, in turbine rotors of ships at the time of turning, rolling or pitching, in propeller-driven planes during turning, etc.

Let us examine gyrostatic action in a gyroscope whose axis together with a bearing frame (Fig. 93) can rotate freely about the horizontal axis OO' of a U-shaped mount. When the mount is set into a forced rotation about the vertical axis as shown in the figure by the $\boldsymbol{\omega}'$ vector, the angular momentum $\mathbf{L}$ of the gyroscope receives during the time interval dt an increment $d\mathbf{L}_1$, a vector directed beyond the plane of the figure. That increment is induced by the moment $\mathbf{M}_1$ of a couple exerted on the gyroscope axis by the frame. The corresponding gyroscopic forces exerted by the gyroscope axis on the frame make the frame turn about the horizontal axis OO'. In the process the vector $\mathbf{L}$ acquires an additional increment $d\mathbf{L}_2$ which is in its turn caused by the moment $\mathbf{M}_2$ of the couple which the frame exerts on the gyroscope axis. Consequently, the gyroscope axis turns so that the vector $\mathbf{L}$ tends to coincide in its direction with the vector $\boldsymbol{\omega}'$.

Thus, during the time interval dt the angular momentum $\mathbf{L}$ of the gyroscope acquires an increment $d\mathbf{L} = d\mathbf{L}_1 + d\mathbf{L}_2 = (\mathbf{M}_1 + \mathbf{M}_2)\, dt$. As this takes place, the frame experiences the gyroscopic couple

$$\mathbf{M}' = -(\mathbf{M}_1 + \mathbf{M}_2).$$

The component of that couple $\mathbf{M}'_1 = -\mathbf{M}_1$ makes the frame turn about the horizontal axis OO', while the other component $\mathbf{M}'_2 = -\mathbf{M}_2$ *opposes* the turning of the whole system about the vertical axis (in contrast to the previous case).

The gyroeffect underlies various applications of gyroscopes: the gyrocompass, gyroscopic stabilizers, the gyrosextant, etc.

Problems to Chapter 5

● 5.1. Find the maximum and minimum distances of the planet A from the Sun S if at a certain moment of time it was at the distance r_0 and travelled with the velocity v_0, with the angle between the radius vector $\mathbf{r}_0$ and the vector $\mathbf{v}_0$ being equal to φ (Fig. 94).

Solution. Let us make use of the laws of conservation of angular momentum and energy. The centre of the Sun is the point relative to which the angular momentum of the planet remains constant. Therefore,

$$r_0 m v_0 \sin \varphi = r m v,$$

where m is the mass of the planet. The angular momentum of the planet at a given moment of time enters into the left-hand side of

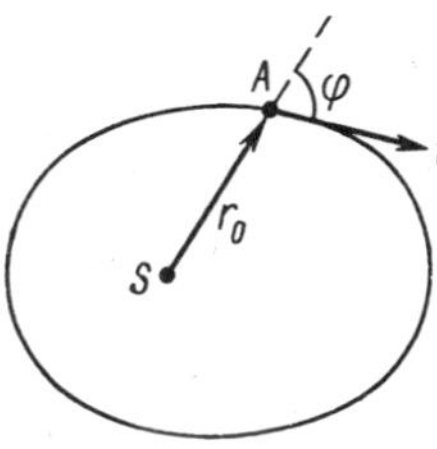

Fig. 94

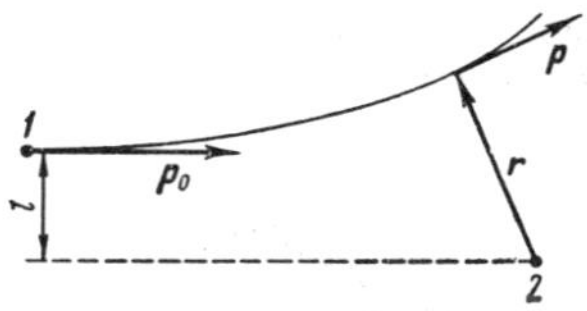

Fig. 95

that equality, while the right-hand side contains its angular momentum at the maximum (minimum) distance r when $\mathbf{r} \perp \mathbf{v}$.

From the energy conservation law it follows that

$$mv_0^2/2 - \gamma mM/r_0 = mv^2/2 - \gamma mM/r,$$

where M is the mass of the Sun and γ is the gravitational constant.

Eliminating v from these two equations, we get

$$r = \frac{r_0}{2-\alpha}\left(1 \pm \sqrt{1-\alpha(2-\alpha)\sin^2\varphi}\right),$$

where $\alpha = r_0 v_0^2/\gamma M$. The plus sign in front of the radical sign corresponds to r_{max} and the minus sign to r_{min}.

● 5.2. Particle *1* located far from particle *2* and possessing the kinetic energy T_0 and mass m_1 strikes particle *2* of mass m_2 through the aiming parameter l, the arm of the momentum vector relative to particle *2* (Fig. 95). Each particle carries a charge $+q$. Find the smallest distance between the particles when

(1) $m_1 \ll m_2$;

(2) m_1 is comparable to m_2.

Solution. 1. The condition $m_1 \ll m_2$ means that in the process of interaction particle *2* is practically motionless. The vector of the force acting on particle *1* passes continuously through the point at which particle *2* is located. Consequently, the angular momentum of particle *1* relative to motionless particle *2* is conserved. Hence,

$$lp_0 = r_{min} p,$$

where the left-hand side represents the angular momentum of particle *1* located far from particle *2* and the right-hand side is the angular momentum of particle *1* at the moment of the closest approach, when $\mathbf{r} \perp \mathbf{p}$ (Fig. 95.) Next, from the energy conservation law it follows that

$$T_0 = T + kq^2/r_{min},$$

where T is the kinetic energy of particle *1* at the moment of the closest approach. Having solved these two equations (and taking into account the relationship between p_0 and T_0), we obtain

$$r_{min} = \frac{kq^2}{2T_0}\left(1 + \sqrt{1 + (2lT_0/kq^2)^2}\right). \quad (1)$$

2. In this case one cannot assume that particle *2* is at rest in the process of interaction. It is advisable to seek the solution here in the C frame, in which the "collision" occurs the way it is shown in Fig. 96. This system of two particles is assumed closed, and therefore its internal angular momentum is conserved:

$$l\tilde{p}_{10} = r_{min}\tilde{p}_1, \quad (2)$$

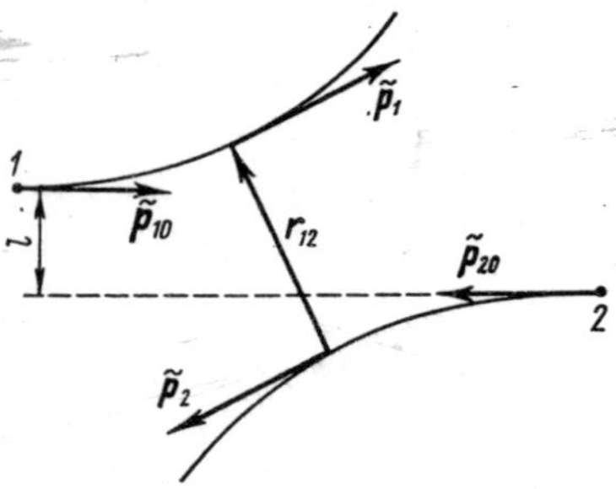

Fig. 96

where account is taken of the fact that $\mathbf{r}_{12} \perp \tilde{\mathbf{p}}_1$ at the moment of the closest approach (see Fig. 95). Moreover, in accordance with the energy conservation law

$$\tilde{T}_0 = \tilde{T} + kq^2/r_{min}, \quad (3)$$

where $\tilde{T}_0$ and $\tilde{T}$ are the total kinetic energies of the particles in the C frame, respectively, at the moment when they are far from each other and at the moment of their closest approach. From Eqs. (2) and (3) we get expression (1), only with $\tilde{T}_0$ substituted for T_0. What is more, in this case (particle *2* is originally at rest)

$$\tilde{T}_0 = \frac{m_2}{m_1 + m_2} T_0,$$

in accordance with Eq. (4.16). Note that if $m_1 \ll m_2$, then $\tilde{T}_0 \approx T_0$ and the expression for r_{min} is completely identical to Eq. (1).

● 5.3. A small sphere is suspended at the point O by means of a light non-stretchable thread of length l. Then the sphere is swung through an angle θ from the vertical and imparted an initial velocity v_0 perpendicular to the vertical plane in which the thread is located. At what velocity v_0 is the maximum swinging angle equal to $\pi/2$?

Solution. The swinging sphere experiences two forces: the gravitational force and the tension of the thread. It is not difficult to see that the moment of these forces $M_z \equiv 0$ relative to the vertical axis z passing through the point O. Consequently, the angular momentum L_z of the sphere relative to the given axis is constant, or

$$l \sin\theta \cdot mv_0 = l \cdot mv, \quad (1)$$

where m is the mass of the sphere, and v is its velocity in the position when the thread forms the angle $\pi/2$ with the vertical.

The sphere moves in the Earth's gravitational field under the influence of an external force, the tension of the thread. That force is always perpendicular to the velocity vector of the sphere and therefore does not perform any work. It follows that in accordance with Eq. (3.32) the mechanical energy of the sphere in the Earth's gravitational field remains constant:

$$mv_0^2/2 = mv^2/2 + mgl\cos\theta, \quad (2)$$

where the right-hand side of the equality corresponds to the horizontal position of the thread.

The simultaneous solution of Eqs. (1) and (2) yields

$$v_0 = \sqrt{2gl/\cos\theta}.$$

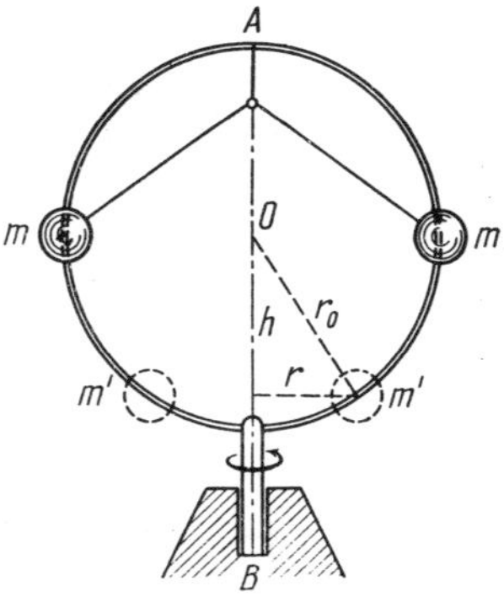

Fig. 97

● 5.4. Two identical small couplings are positioned on a rigid wire ring of radius r_0 which can freely rotate about the vertical axis AB (Fig. 97). They are joined together by a thread and set in the position m-m. Then the whole assembly is set into rotation with the angular velocity ω_0, whereupon the thread is burnt at the point A. Assuming the mass of the assembly to be concentrated mainly in the couplings, find its angular velocity at the moment when the couplings have slid down (without friction) to the extreme lower position m'-m'.

Solution. Suppose that in the lower position the couplings are located at the distance r from the rotation axis and the angular velocity of the assembly is ω. Then, from the laws of conservation of energy and angular momentum relative to the rotation axis, it follows that

$$r^2\omega^2 - r_0^2\omega_0^2 = 2gh; \qquad r^2\omega = r_0^2\omega_0,$$

where h is the difference in the height of the couplings in the upper and lower positions. Besides, from Fig. 97 it is seen that

$$r_0^2 = r^2 + h^2.$$

These three equations, when solved simultaneously, yield

$$\omega = \left(1 + \sqrt{1 + (4g/r_0\omega_0^2)^2}\right)\omega_0/2.$$

● 5.5. A smooth rod rotates freely in a horizontal plane with the angular velocity ω_0 about a stationary vertical axis O (Fig. 98) relative to which the rod's moment of inertia is equal to I. A small coupling of mass m is located on the rod close to the rotation axis and is tied to it by a thread. When the thread is burned, the coupling starts sliding along the rod. Find the velocity v' of the coupling relative to the rod as a function of its distance r from the rotation axis.

Solution. In the process of motion of the given system the kinetic energy and the angular momentum relative to the rotation axis do not vary. Hence, it follows that

$$I\omega_0^2 = I\omega^2 + mv^2; \quad I\omega_0 = (I + mr^2)\,\omega,$$

where $v^2 = v'^2 + \omega^2 r^2$ (Fig. 98). From these equations we obtain

$$v' = \omega_0 r / \sqrt{1 + mr^2/I}.$$

● **5.6.** A bullet A flying horizontally hits and remains in a vertical uniform rod of mass m and length l_0 hinged by its upper end at the

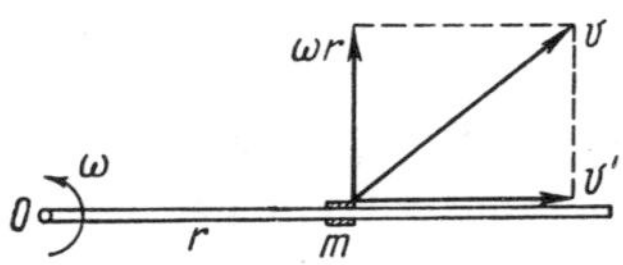

Fig. 98

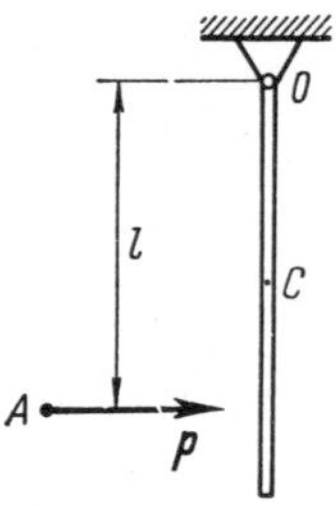

Fig. 99

point O (Fig. 99). The bullet has the momentum $\mathbf{p}$ and hits the rod at a point lying at the distance l from the point O. Disregarding the mass of the bullet, find

(1) the momentum increment of the bullet-rod system during the time of motion of the bullet in the rod;

(2) the angular velocity acquired by the rod with regard to the internal angular momentum of the bullet, which is equal to $\tilde{\mathbf{L}}$ and coincides in direction with the vector $\mathbf{p}$ (the bullet spins about its motion direction).

Solution. 1. The bullet-rod system is non-closed: apart from the counterbalancing forces a horizontal component of the reaction force appears in the process of motion of the bullet at the point O of the rod. That component brings about the momentum increment of the system:

$$\Delta p = mv_C - p,$$

where v_C is the velocity of the centre of inertia of the rod after the bullet has hit.

Since all external forces in this process pass through the point O, the angular momentum of the system remains constant relative to any axis passing through that point as long as the bullet moves in the rod. Choosing the axis that is at right angles to the figure plane, we write

$$lp = I\omega,$$

where I is the moment of inertia of the rod relative to the axis thus chosen, and ω is the angular velocity of the rod immediately after the bullet has stopped in it.

With allowance made for $v_C = \omega r$, where r is the distance from the point O to the centre of inertia of the rod, these two equations yield

$$\Delta p = (3l/2l_0 - 1)\,p.$$

It is seen that the sign of the increment Δp depends on the ratio l/l_0. Specifically, when $l/l_0 = 2/3$, $\Delta p = 0$, i.e. the momentum of the system does not vary as long as the bullet keeps moving in the rod. This signifies that in this case the horizontal component of the reaction force at the point O is absent.

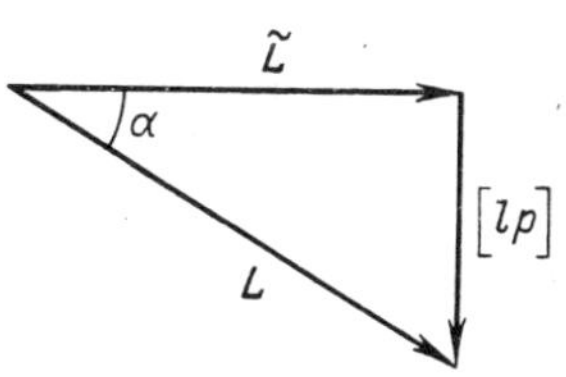

Fig. 100

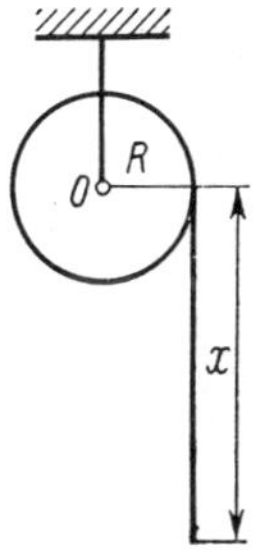

Fig. 101

2. In this case the angular momentum of the system with respect to the point O also remains constant as long as the bullet moves in the rod, and consequently in accordance with Eq. (5.23)

$$\tilde{\mathbf{L}} + [\mathbf{lp}] = \mathbf{L}.$$

Here the angular momentum of the bullet relative to the point O is written on the left-hand side and on the right-hand side we wrote the angular momentum of the rod (together with the bullet stuck in it) immediately after the bullet stopped (see Fig. 100 where all three vectors are located in the horizontal plane).

Let us calculate the vector $\mathbf{L}$ when the rod (with the bullet) acquires the angular velocity $\boldsymbol{\omega}$. We shall consider a small element of the rod of mass dm located at the distance $\mathbf{r}$ from the point O. Its angular momentum relative to the point O is equal to

$$d\mathbf{L} = [\mathbf{r},\ dm\mathbf{v}] = dm \cdot r^2 \boldsymbol{\omega} = (m\boldsymbol{\omega}/l_0)\, r^2\, dr,$$

where $\mathbf{v}$ is the velocity of the given element. Integrating this expression over all elements, we obtain

$$\mathbf{L} = ml_0^2\boldsymbol{\omega}/3.$$

Thus,

$$\tilde{\mathbf{L}} + [\mathbf{lp}] = ml_0^2\boldsymbol{\omega}/3.$$

Using this formula, we obtain in accordance with Fig. 100

$$\omega = 3\sqrt{\tilde{L}^2 + l^2p^2}/ml_0^2.$$

From the same figure one can find also the direction of the vector $\boldsymbol{\omega}$ (the angle α).

● **5.7.** A uniform solid cylinder of mass m_0 and radius R can rotate without friction about a stationary horizontal axis O (Fig. 101). A thin non-stretchable cord of length l and mass m is wound in one layer over the cylinder. Find the angular acceleration of the cylinder as a function of the length x of the overhanging piece of cord. Slipping is assumed absent and the centre of gravity of the wound portion of cord is supposed to be located at the cylinder's axis.

Solution. Let us make use of the equation of moments (5.15) relative to the O axis. For this purpose we find the angular momentum of the system L_z relative to the given axis and the corresponding force moment M_z. The angular momentum is

$$L_z = I\omega_z + Rmv = (m_0/2 + m)\, R^2\omega_z,$$

where allowance is made for the moment of inertia of the cylinder $I = m_0R^2/2$ and $v = \omega_zR$ (no cord slipping). The moment of the external gravitational forces relative to the O axis is

$$M_z = Rmgx/l.$$

Differentiating L_z with respect to time and substituting dL_z/dt and M_z into the equation of moments, we obtain

$$\beta_z = 2mgx/lR\,(m_0 + 2m).$$

● **5.8.** A uniform disc of radius r_0 lies on a smooth horizontal plane. A similar disc spinning with the angular velocity ω_0 is carefully lowered onto the first disc. How soon do both discs spin with the same angular velocity if the friction coefficient between them is equal to k?

Solution. First, let us find the steady-state rotation angular velocity ω. From the law of conservation of angular momentum it follows that $I\omega_0 = 2I\omega$, where I is the moment of inertia of each disc relative to the common rotation axis. Hence,

$$\omega = \omega_0/2.$$

Now let us examine the behaviour of one of the discs, for example, the lower one. Its angular velocity varies due to the moment M of the friction forces. To calculate M, we single out on the upper surface plane of the disc an elementary ring with radii r and $r + dr$.

The moment of the friction forces acting on the given ring is equal to

$$dM = rk\,(mg/\pi r_0^2)\,2\pi r\,dr = (2kmg/r_0^2)\,r^2\,dr,$$

where m is the mass of each disc. Integrating this expression with respect to r between 0 and r_0, we get

$$M = (2/3)\,kmgr_0.$$

In accordance with Eq. (5.30), the angular velocity of the lower disc increases by $d\omega$ over the time interval

$$dt = (I/M)\,d\omega = (3r_0/4\,kg)\,d\omega.$$

Integrating this equation with respect to ω between 0 and $\omega_0/2$, we find the sought time

$$t = 3r_0\omega_0/8kg.$$

● 5.9. A uniform cylinder is placed on a horizontal board (Fig. 102). The coefficient of friction between them is equal to k. The board is

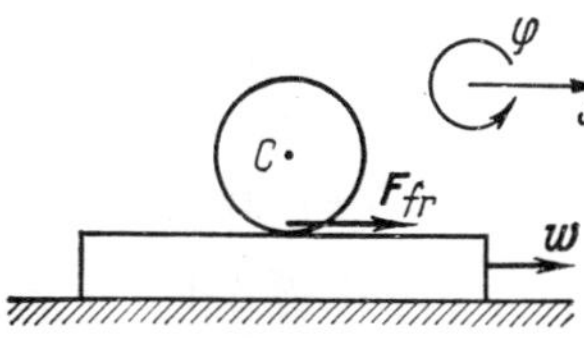

Fig. 102

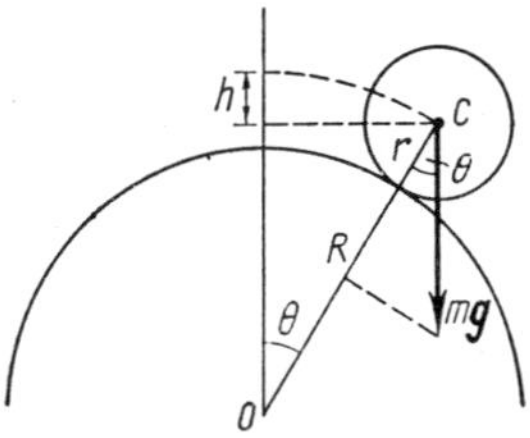

Fig. 103

imparted a constant acceleration w in a horizontal direction at right angles to the cylinder's axis. Find

(1) the acceleration of the cylinder axis in the absence of slipping;

(2) the limiting value w_{lim} for which there is no slipping.

Solution. 1. Choosing the positive directions for x and φ as shown in Fig. 102, we write the equation of motion of the cylinder axis and the equation of moments in the C frame relative to that axis:

$$mw_C = F_{fr}; \quad I\beta = rF_{fr},$$

where m and I are the mass and the moment of inertia of the cylinder relative to its axis. In addition, the condition for the absence of slipping of the cylinder yields the kinematic relation of the accelerations:

$$w - w_C = \beta r.$$

From these three equations we obtain $w_C = w/3$.

2. Let us find from the previous equations the magnitude of the friction force F_{fr} ensuring that the cylinder rolls without slipping: $F_{fr} = mw/3$. The maximum possible value of that force is equal

to kmg. Hence,

$$w_{lim} = 3kg.$$

● **5.10.** A uniform sphere of radius r starts rolling down without slipping from the top of another sphere of radius R (Fig. 103). Find the angular velocity of the sphere after it leaves the surface of the other sphere.

Solution. First of all note that the angular velocity of the sphere after it leaves does not change. Therefore the problem reduces to the determination of its magnitude at the moment of breaking-off.

Let us write the equation of motion for the centre of the sphere at the moment of breaking-off:

$$mv^2/(R + r) = mg \cos\theta,$$

where v is the velocity of the centre of the sphere at that moment, and θ is the corresponding angle (Fig. 103). The velocity v can be found from the energy conservation law:

$$mgh = mv^2/2 + I\omega^2/2,$$

where I is the moment of inertia of the sphere relative to the axis passing through the sphere's centre. In addition,

$$v = \omega r; \quad h = (R + r)(1 - \cos\theta).$$

From these four equations we obtain

$$\omega = \sqrt{10g(R+r)/17r^2}.$$

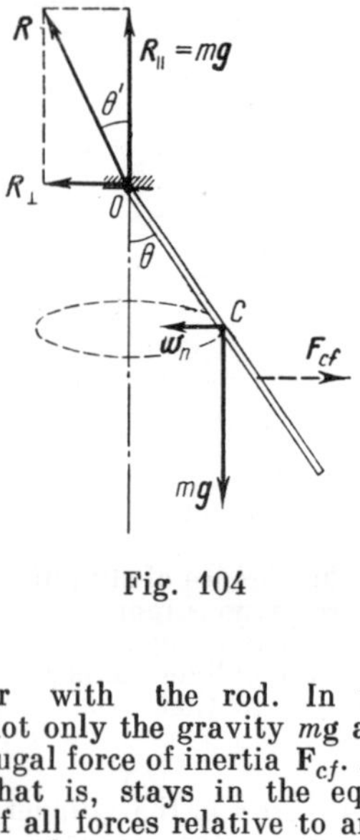

Fig. 104

● **5.11.** A thin uniform rod of mass m and length l rotates with the constant angular velocity ω about the vertical axis passing through the rod's suspension point O (Fig. 104). In so doing, the rod describes a conical surface with a half-aperture angle θ. Find the angle θ as well as the magnitude and direction of the reaction force **R** at the point O.

Solution. Let us consider the frame rotating about the vertical axis together with the rod. In this reference frame the rod experiences not only the gravity mg and the reaction force **R** but also the centrifugal force of inertia $\mathbf{F}_{cf}$. As the rod rests in the given reference frame, that is, stays in the equilibrium position, the resultant moment of all forces relative to any point and the resultant of all forces are equal to zero.

It is only gravity and the centrifugal forces of inertia that produce a moment relative to the point O. From the equality of the moments of these forces it follows that

$$(mgl/2)\sin\theta = M_{cf}. \tag{1}$$

Let us calculate M_{cf}. The elementary moment of the force of inertia that acts on the rod element dx located at the distance x from the point O is equal to

$$dM_{cf} = (m\omega^2/l) \sin\theta \cos\theta x^2\, dx.$$

Integrating this expression over the whole length of the rod, we get

$$M_{cf} = (m\omega^2 l^2/3) \sin\theta \cos\theta. \tag{2}$$

It follows from Eqs. (1) and (2) that

$$\cos\theta = 3g/2\omega^2 l. \tag{3}$$

Now let us determine the magnitude and direction of the vector **R**. In the reference frame where the rod rotates with the angular velocity ω its centre of inertia, the point C, moves along a horizontal circle. Consequently, from the law of motion of a centre of inertia, that is, Eq. (4.11), it immediately follows that the vertical component of the vector **R** is $R_{\parallel} = mg$, while the horizontal component $R_{\perp}$ is determined from the equation $mw_n = R_{\perp}$, where w_n is the normal acceleration of the centre of inertia C. Hence

$$R_{\perp} = (m\omega^2 l/2) \sin\theta. \tag{4}$$

The magnitude of the vector **R** is equal to

$$R = \sqrt{(mg)^2 + R_{\perp}^2} = (m\omega^2 l/2) \sqrt{1 + 7g^2/(4\omega^4 l^2)},$$

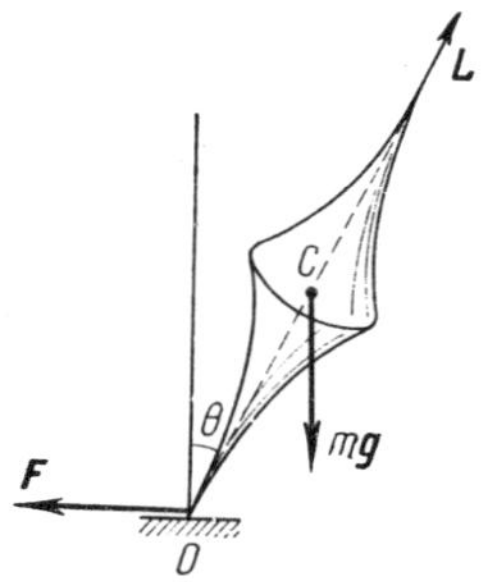

Fig. 105

and its direction, specified by the angle θ' between the vector **R** and the vertical, is determined from the formula $\cos\theta' = mg/R$. It is interesting to note that $\theta' \neq \theta$, i.e. the vector **R** does not coincide with the rod in direction. One can easily make sure of this by expressing $\cos\theta'$ via $\cos\theta$:

$$\cos\theta' = 4\cos\theta / \sqrt{9 + 7\cos^2\theta}.$$

It is seen from this that $\cos\theta' > \cos\theta$, i.e. $\theta' < \theta$, as it is in fact shown in Fig. 104.

Note that the equivalent of the forces of inertia $\mathbf{F}_{cf}$ passes not through the point C but below it. Indeed, $F_{cf} = R_{\perp}$ and is determined by Eq. (4), whereas the resultant moment M_{cf} is determined by Eq. (2). It follows from these formulae that the arm of the vector $\mathbf{F}_{cf}$ relative to the point O is equal to $(2l/3)\cos\theta$ (Fig. 104).

● **5.12.** A spinning top of mass m whose axis forms an angle θ with the vertical precesses about the vertical axis passing through the point of support O. The angular momentum of the top is equal to L, and its centre of inertia is located at the distance l from the point O. Find the magnitude and direction of the vector **F** which is the horizontal component of the reaction force at the point O.

Solution. In accordance with Eq. (5.38) the angular velocity of precession is

$$\omega' = mgl/L.$$

Since the centre of inertia of the top moves along the circle, the vector **F** is oriented as illustrated in Fig. 105 (this vector precesses together with the axis of the top).

From the motion equation for the centre of inertia we obtain

$$m\omega'^2 l \sin\theta = F.$$

And finally,

$$F = (m^3 g^2 l^3/L^2) \sin\theta.$$

It should be pointed out that if the point of support of the top were located on an ideally smooth plane, the top would precess with the same angular velocity but about the vertical axis passing through the centre of the top, the point C in Fig. 105.

PART TWO

RELATIVISTIC MECHANICS

CHAPTER 6

KINEMATICS IN THE SPECIAL THEORY OF RELATIVITY

§ 6.1. Introduction

The special theory of relativity proposed by Einstein in 1905 called for a review of all concepts of classical physics and primarily the concepts of time and space. Therefore, this theory, in accordance with its basic contents, can be referred to as a *physical* study of time and space. The study is called physical because the properties of space and time are analysed in this theory in close connection with the laws governing physical phenomena. The term "special" implies that this theory considers phenomena only in inertial reference frames.

We shall begin this section with a brief review of prerelativistic physics, dwelling in particular on the problems that led to the appearance of the theory of relativity.

Basic notions of prerelativistic physics. First, we shall recall those notions of space and time that are associated with Newton's laws, i.e. that underlie classical mechanics.

1. Space, which has three dimensions, obeys Euclidean geometry.

2. Together with three-dimensional space and independent of it, there exists time. Time is independent in the same sense that the three dimensions are independent of each other. But for all that, time relates to space through the laws of motion. Specifically, time is measured by a clock, which is basically an instrument utilizing one or another periodic process providing a time scale. Therefore, it is impossible to determine time irrespective of some periodic process, i.e. irrespective of motion.

3. Dimensions of solid bodies (scales) and time intervals between given events are identical in different reference frames. This corresponds to the Newtonian concept of absolute space and time, according to which their properties are assumed to be independent of the reference frame, that is, space and time are the same for all reference frames.

4. The Galileo-Newton law of inertia is assumed to be valid, according to which a body experiencing no influence from other bodies moves rectilinearly and uniformly. This law maintains the existence of inertial reference frames in which Newton's laws hold true (as well as the Galilean principle of relativity).

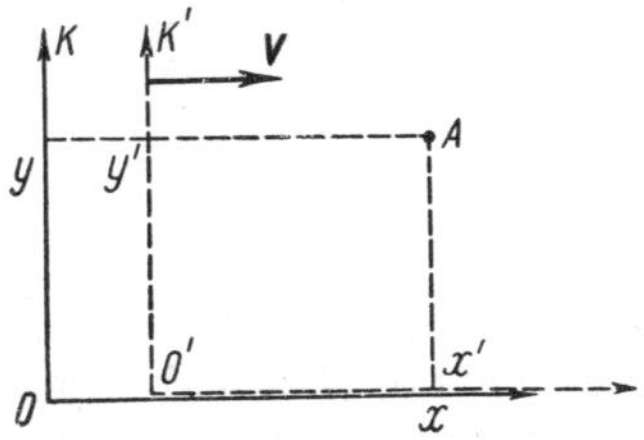

Fig. 106

5. From the above notions the Galilean transformation follows, expressing the space-time relation of any event in different inertial reference frames. If the reference frame K' moves relative to the K frame with the velocity $\mathbf{V}$ (Fig. 106) and the zero time reading corresponds to the moment when the origins O' and O of the two frames coincide, then*

$$x' = x - Vt; \quad y' = y; \quad t' = t. \tag{6.1}$$

From this it follows that the coordinates of any event are relative, i.e. have different values in different reference frames; the moment of time at which an event occurs is however the same in different frames. This testifies to the fact that time flows identically in different reference frames. That seemed to be so obvious that it was not even stated as a special postulate.

From Eq. (6.1) the classical law of velocity transformation (composition) follows immediately:

$$\mathbf{v}' = \mathbf{v} - \mathbf{V}, \tag{6.2}$$

where $\mathbf{v}'$ and $\mathbf{v}$ are the velocities of a point (particle) in the K' and K frames respectively.

* Hereafter we shall limit ourselves to only two spatial coordinates x and y. The z coordinate behaves as y in all respects.

6. The Galilean principle of relativity holds: all inertial reference frames are equivalent in terms of mechanics, all laws of mechanics are identical in these reference frames, or, in other words, are invariant relative to the Galilean transformation.

7. The principle of long-range action is valid: interactions between bodies propagate instantaneously, i.e. with an infinitely high velocity.

These notions of classical mechanics were in complete accord with the totality of experimental data available at that time (it should be noted, though, that those data related to the study of bodies moving with velocities much lower than the velocity of light). The validity of these notions was confirmed by the very successful development of mechanics itself. Therefore, the notions of classical mechanics about the properties of space and time were thought to be so fundamental as not to raise any doubts about their truth.

The first to be put to the test was the Galilean principle of relativity, which is known to be applicable only in mechanics, the only division of physics advanced sufficiently by that time. As other branches of physics, in particular, optics and electrodynamics, were developing, the natural question arose: does the principle of relativity cover other phenomena as well? If not, then using these (non-mechanical) phenomena one can in principle distinguish inertial reference frames and try to find a primary, or absolute, reference frame.

One of such phenomena that was expected to occur differently in different reference frames is the propagation of light. In accordance with the predominant wave theory of that time waves of light must propagate with a certain velocity relative to a certain hypothetical medium ("luminiferous ether") whose nature, however, was debated among scientists. Still, whatever the nature of that medium, it surely cannot rest in all inertial reference frames at once. Consequently, one can distinguish one inertial frame, the absolute frame, which is stationary with respect to the "luminiferous ether". It was supposed that in that (and only in that) reference frame light propagates with the equal velocity c in all directions. If a certain inertial reference frame moves with the velocity $\mathbf{V}$ relative to the ether, the velocity of light $\mathbf{c}'$ in that reference frame must obey the conventional law of velocity composition (6.2), i.e. $\mathbf{c}' = \mathbf{c} - \mathbf{V}$. This assumption was tested experimentally by Michelson (together with Morley).

Michelson's experiment. The purpose of this experiment was to detect the "true" motion of the Earth relative to the ether. Michelson in his experiment took advantage of the motion of the Earth along its orbit with the velocity 30 km/s. The idea of the experiment was the following.

The light from the source S (Fig. 107) was emitted in two mutually perpendicular directions, reflected from the mirrors A and B located at the same distance l from the source S and finally returned to the point S. In this experiment a comparison was made between the time taken by light to cover the path SAS and the time taken to cover the path SBS.

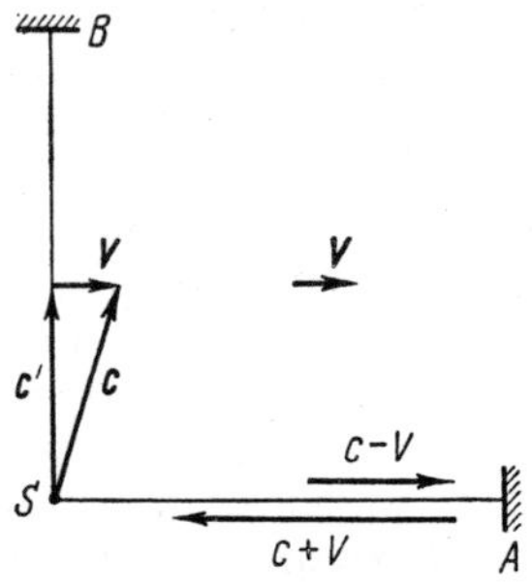

Fig. 107

Let us suppose that at the moment of the experiment the set-up moves together with the Earth so that its velocity $\mathbf{V}$ relative to the ether is directed along SA. If the velocity of light obeys the conventional law of velocity composition (6.2), light moves along the path SA with the velocity $c - V$ relative to the set-up (the Earth) and in the reverse direction with the velocity $c + V$. Then the time spent by light to traverse the path SAS is equal to

$$t_{\parallel} = \frac{l}{c-V} + \frac{l}{c+V} = \frac{2l}{c} \frac{1}{1-(V/c)^2}.$$

Along the path SBS the velocity of light relative to the set-up is equal to $c' = \sqrt{c^2 - V^2}$ (Fig. 107) and the time taken to cover that path is equal to

$$t_{\perp} = \frac{2l}{\sqrt{c^2 - V^2}} = \frac{2l}{c} \frac{1}{\sqrt{1-(V/c)^2}}.$$

Comparing the expressions for $t_{\parallel}$ and $t_{\perp}$, we see that light must take different time periods to cover these paths. By measuring the time difference $t_{\parallel} - t_{\perp}$ the velocity of the set-up (the Earth) relative to the ether can be determined.

Despite the fact that the time difference was expected to be extremely small, the set-up was capable of observing that difference using a very sensitive interferometer technique.

For all that, the result of the experiment proved to be negative: a time difference was not detected. Surely, by sheer accident, the Earth could happen to be motionless relative to the ether at the time when the experiment was conducted. But then in half a year, for example, the Earth's velocity would have reached 60 km/s. The repetition of the experiment in half a year, however, did not bring the result expected.

More accurate experiments of the same kind performed later corroborated the original result.

The negative result of Michelson's experiment contradicted what was expected from the Galilean transformation (velocity composition). It also showed that motion relative to the ether cannot be detected and the velocity of light is independent of the motion of a light source since its motion with respect to the ether is different at different seasons of the year.

Some astronomical observations (e.g. of double stars) also point to the fact that the velocity of light does not depend on the velocity of a source. A number of special experiments carried out later gave the same evidence.

By the beginning of the twentieth century theoretical and experimental physics faced a serious challenge. On the one hand, the theory predicted various effects permitting the principal (absolute) reference frame to be distinguished from the great number of inertial frames. On the other hand, persistent attempts to detect these effects experimentally inevitably terminated in failure. The experiment perfectly confirmed the validity of the principle of relativity for all phenomena, including those which were thought incompatible with that principle.

A few attempts were ventured to explain the negative outcome of the Michelson experiment and some other simi-

13—0539

lar experiments in terms of classical mechanics. However, all of them turned out to be unsatisfactory in the final analysis. The cardinal solution of this problem was provided only in Einstein's theory of relativity.

§ 6.2. Einstein's Postulates

A profound analysis of all experimental and theoretical data accumulated by the beginning of the twentieth century led Einstein to review the initial notions of classical physics and primarily the concepts of space and time. As a result, he created the special theory of relativity, which proved to be a logical completion of classical physics.

This theory adopts unaltered such concepts of classical mechanics as Euclidean space and the Galileo-Newton law of inertia. As to the statement concerning the constancy of size of solid bodies and of time intervals in different reference frames, Einstein noticed that these representations emerged from the observations of bodies moving with low velocities, and therefore their extrapolation to higher velocities is unwarranted and, for this reason, incorrect. Only experiment can give evidence concerning the true properties of space and time. The same can be said about the Galilean transformation and the principle of long-range action.

Einstein proposed two postulates, or principles as the foundation of the special theory of relativity, which were backed up by experimental data (and primarily by the Michelson experiment):

(1) the principle of relativity,

(2) independence of the velocity of light of the velocity of a source.

The first postulate is a generalization of the Galilean principle of relativity to cover all physical processes: *all physical phenomena proceed identically in all inertial reference frames*; *all laws of nature and the equations describing them are invariant*, *i.e. keep their form on transition from one inertial reference frame to another*. In other words, *all inertial reference frames are equivalent* (indiscernible) *in their physical properties*; basically, no experiment whatever can distinguish one of the frames as preferable.

The second postulate states that *the velocity of light in vacuo is independent of the motion of a light source and is the same in all directions.* This means that *the velocity of light in vacuo is the same in all inertial reference frames.* Thus, the velocity of light holds a most unique position. In contrast to all other velocities, which change on transition from one reference frame to another, the velocity of light *in vacuo* is an invariant quantity. As we shall see later, the existence of such a velocity essentially modifies the notions of space and time.

It also follows from Einstein's postulates that light propagates *in vacuo* at the *ultimate* velocity: no other signal, no interaction between any two bodies can propagate with a velocity exceeding that of light *in vacuo.* It is precisely due to its limiting nature that the velocity of light is the same in all reference frames. Indeed, in accordance with the principle of relativity the laws of nature must be identical in all inertial reference frames. The fact that the velocity of any signal cannot exceed the ultimate value is also a law of nature. Consequently, the ultimate velocity value, the velocity of light *in vacuo*, must be the same in all inertial reference frames.

Specifically, the existence of an ultimate velocity presupposes the limiting of velocities of moving particles by the value c. If otherwise, particles could transmit signals (or interactions between bodies) with a velocity exceeding the ultimate one. Thus, in accordance with Einstein's postulates all possible velocities of moving bodies and of interaction propagation are limited by the value c. Thus, the principle of long-range action of classical mechanics does not hold any more.

The whole content of the special theory of relativity follows from these two postulates. By the present time, both Einstein's postulates, as well as all of their consequences, have been convincingly confirmed by the totality of experimental data accumulated so far.

Clock synchronization. Prior to drawing any conclusions from these postulates Einstein carefully analysed the methods of measuring space and time. First of all, he noticed that neither a point of space nor a time moment at which a certain event occurs possesses a physical reality; it is the *event* itself that does. To describe an event in a given reference frame, one must indicate the point and the moment of time at which it occurs.

The location of the point at which the event occurs can be determined in terms of Euclidean geometry by means of rigid scales and expressed in Cartesian coordinates. In this case classical mechanics resorted to quite workable methods of comparing quantities being measured against reference standards.

The corresponding moment of time can be determined by means of a clock placed at the point of the reference frame where the given event occurs. This method, however, is not satisfactory any more when we have to compare events occurring at *different* points, or, which is the same, to intercompare the moments of time of events happening at points removed from the clock.

Indeed, to compare the moments of time (time readings) at different points of a reference frame one has to define first what is the *universal* time for all points of the reference frame. In other words, we have to ensure a *synchronous* rate of all clocks of the given reference frame.

It is clear that the synchronization of clocks positioned at different points of the reference frame can be accomplished only by means of some signals. The fastest signals suitable for the purpose are light and radio signals propagating with the known velocity c. The choice of these signals is also stipulated by the fact that their velocity is independent of the direction in space and is the same in all inertial reference frames.

Next, we can do as follows. An observer located, for example, at the origin O of a given reference frame broadcasts a time signal at the moment t_0 by his clock. At the moment when this signal reaches the clock located at the known distance r from the point O the clock is set so that it registers $t = t_0 + r/c$, i.e. the signal delay is taken into account. The repetition of signals after definite time intervals permits all observers to synchronize the rate of their clocks with that of the clock at the point O. This operation having been performed, one can claim that all the clocks of the given reference frame register the same time at each moment.

It is essential to point out that time defined that way refers only to the reference frame relative to which the synchronized clocks are motionless.

Relationships between events. Let us investigate the spatial and temporal relationships between given events in different inertial reference frames.

Even in classical mechanics the spatial relationships between different events depend on the reference frame to which they belong. For example, two consecutive flashes in a moving train occur at the same point of a reference frame fixed to the train but at different points of a reference frame fixed to the railroad bed. The statement that alternative events occur at the same point or at a certain distance between them makes sense only if the reference frame to which that statement refers is indicated.

Fig. 108

By contrast, in classical mechanics temporal relationships between events are assumed independent of the reference frame. This means that if two events occur simultaneously in one reference frame they are simultaneous in all other frames. Generally, the time interval between the given events is assumed to be the same in all frames.

However, it is easy to see that this is not actually the case: *simultaneity* (and therefore the rate of time flow) *is a relative notion* that makes sense only when the reference frame to which that notion relates is indicated. By means of simple reasoning we shall illustrate how two events simultaneous in one reference frame prove to occur at different time moments in another reference frame.

Imagine a rod AB moving with a constant velocity V relative to the reference frame K. A flashbulb is located at the middle point O of the rod, and photodetectors at the ends A and B (Fig. 108). Suppose that at a certain moment the bulb O flashes. Since the velocity with which light propagates in the reference frame fixed to the rod (as in any inertial reference frame) is equal to c in both directions, the light pulses reach the equidistant photodetectors A and B at the same moment of time (in the reference frame "rod") and both photodetectors respond simultaneously.

Things are different in the K frame. In that reference frame the velocity of light pulses propagating in both directions is also equal to c, whereas the paths traversed by them are different. In fact, as the light signals propagate toward the points A and B, the latter shift to the right, and therefore the photodetector A responds earlier than the photodetector B.

Thus, events that are simultaneous in one reference frame are not simultaneous in another one, i.e. in contrast to classical mechanics simultaneity here is a relative notion. This, in turn, means that time flows differently in different reference frames.

If we had at our disposal signals that propagate instantaneously, events simultaneous in one reference frame would also be simultaneous in any other reference frame. This directly follows from the example just considered. In this case the rate of time flow would be independent of the reference frame, and we could talk about the existence of the absolute time that appears in the Galilean transformation. Thus, the Galilean transformation is, in fact, based on the assumption that clock synchronization is accomplished by means of signals propagating instantaneously. However, such signals do not exist.

§ 6.3. Dilation of Time and Contraction of Length

In this section we shall examine three important consequences of Einstein's postulates: the equality of transverse dimensions of bodies in different reference frames, the dilation of the rate of moving clocks, and the contraction of longitudinal dimensions of moving bodies. In § 6.4 we shall generalize the results obtained in the form of the pertinent formulae for transformation of coordinates and time.

Prior to solving these problems we recall that a reference frame is understood as a reference body to which a coordinate grid is fixed together with a number of identical synchronized stationary clocks. Next, the coordinate grids and clocks are assumed to be calibrated in a *like* manner in all inertial reference frames. Clearly, this can be accomplished only by means of length and time standards also similarly realized in all reference frames.

For this purpose we can utilize some natural periodic process providing a natural scale for both length and time measurements, e.g. monochromatic waves emitted by individual atoms resting in a given reference frame. Then in that reference frame the wavelength can serve as a length standard and the corresponding period of oscillation as a time standard. Using these standards we can construct a standard representing *one metre* as a definite number of the given wavelengths and a standard representing *one second* as a definite number of periods of the given oscillations (it should be pointed out that this method has been realized).

A similar technique can be utilized in every inertial reference frame, using the same monochromatic wave emitted by the same atoms resting in each of these reference frames. The method is justified, for in accordance with the principle of relativity physical properties of stationary atoms are independent of the inertial reference frame in which these atoms rest.

Having effected length and time standards in each reference frame, we can move on to solving the fundamental problem of comparing these standards in different reference frames, or, in other words, the comparison of dimensions of bodies and rates of time flow in these frames.

Equality of transverse dimensions of bodies. Let us begin with the comparison of transverse dimensions of bodies in different inertial reference frames. Imagine two inertial reference frames K and K' whose y and y' axes are parallel to each other and perpendicular to the direction of motion of one frame relative to the other (Fig. 109), with the origin O' of the K' frame moving along the straight line passing through the origin O of the K frame. Let us position the rods OA and $O'A'$, which are the metre standards in each of these frames, along the y and y' axes. Next, imagine that at the moment when the y and y' axes coincide the upper end of the left rod makes a marking on the y axis of the K frame. Does that marking coincide with the point A, the upper end of the right rod?

The principle of relativity makes it possible to answer that question: yes, it does. Were it not so, one of the rods would be, for example, shorter than the other from the viewpoint of *both* reference frames, and therefore there would appear an experimental opportunity to distinguish

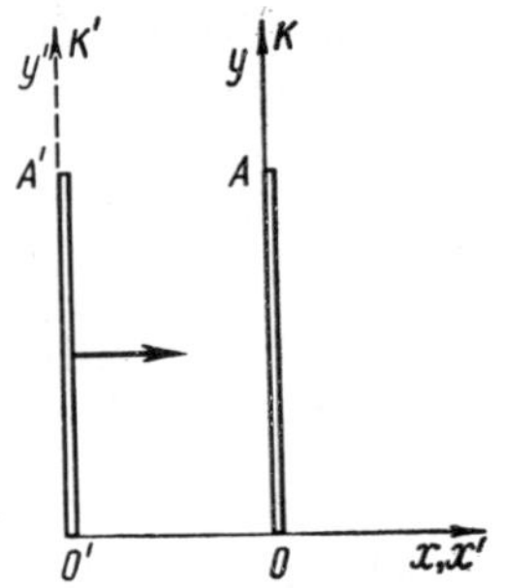

Fig. 109

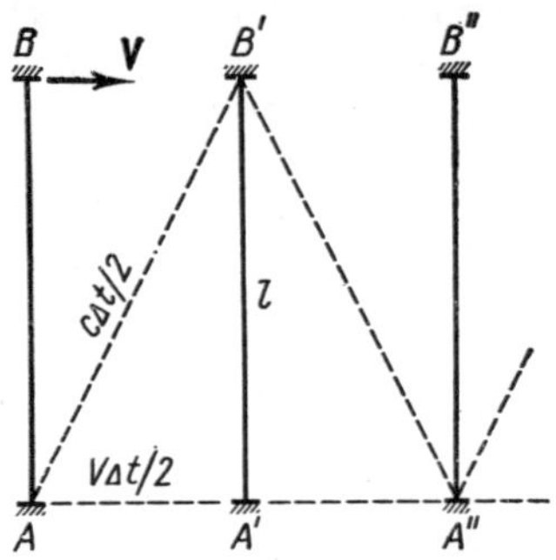

Fig. 110

one inertial reference frame from another by shorter transverse dimensions. This, however, contradicts the principle of relativity.

From this it follows that the transverse dimensions of bodies are the same in all inertial reference frames. This also means that if the origins of the K and K' frames are chosen as indicated, the y' and y coordinates of any point or event coincide, i.e.

$$y' = y. \tag{6.3}$$

This relation represents one of the sought transformations of coordinates.

Dilation of time. Our next task is to compare the rate of time flow in different inertial reference frames. As we already mentioned, time is measured by a clock, which may be any device in which one or another periodic process is used. Accordingly, the theory of relativity customarily deals with the comparison of rates of identical clocks in different inertial reference frames.

The task is easiest when solved by means of the following thought (i.e. basically feasible) experiment. Let us make use of a so-called *light clock*, a rod with mirrors at the ends between which a short light pulse travels back and forth. The period of such a clock is equal to the time interval between two consecutive arrivals of a light pulse at a given end of the rod.

Next, imagine two inertial reference frames K' and K moving relative to each other with the velocity V. Let the light clock AB be at rest in the K' frame and oriented perpendicular to the direction of its motion with respect to the K frame (Fig. 110). Let us see what the rate of the clock is in each respective reference frame K' and K.

In the K' frame the clock is at rest, and its period is

$$\Delta t_0 = 2l/c,$$

where l is the distance between the mirrors.

In the K frame, relative to which the clock moves, the distance between the mirrors is also equal to l, for the transverse dimensions of bodies are the same in different inertial reference frames. However, the path of the light pulse in that reference frame is different (the zigzag of Fig. 110): as the light pulse travels from the bottom mirror toward the upper one, the latter shifts somewhat to the right, etc. Consequently, to get back to the bottom mirror, the light pulse has to cover a longer distance in the K frame while travelling with the same velocity c. Therefore, the light pulse takes a *longer* time to cover that distance as compared to the case when the clock is motionless. Accordingly, the period of the moving clock is longer, i.e. in terms of the K reference frame the clock's rate is *slower*.

Let us designate the period of the moving clock in the K frame by Δt. It follows from the right triangle $AB'A'$ (Fig. 110) that $l^2 + (V\Delta t/2)^2 = (c\Delta t/2)^2$, whence

$$\Delta t = (2l/c)/\sqrt{1-(V/c)^2}.$$

But since $2l/c = \Delta t_0$, then

$$\boxed{\Delta t = \frac{\Delta t_0}{\sqrt{1-\beta^2}},} \qquad (6.4)$$

where $\beta = V/c$ and V is the velocity of the clock in the K frame.

This formula shows that $\Delta t > \Delta t_0$, i.e. the same clock has different rates in different inertial frames: the rate is slower in a reference frame relative to which the clock moves, as compared to the reference frame in which the clock is at rest. In other words, *a moving clock goes slower than a stationary one*. This phenomenon is referred to as the *dilation of time*.

The time measured by the clock moving together with the body in which a certain process takes place is called the *proper time* of that body. It is denoted by Δt_0. As it follows from Eq. (6.4) the proper time interval is the shortest. The time duration Δt of the same process in another reference frame depends on the velocity V of that frame with respect to the body in which the process takes place. This dependence is quite appreciable at velocities V comparable to that of light (Fig. 111).

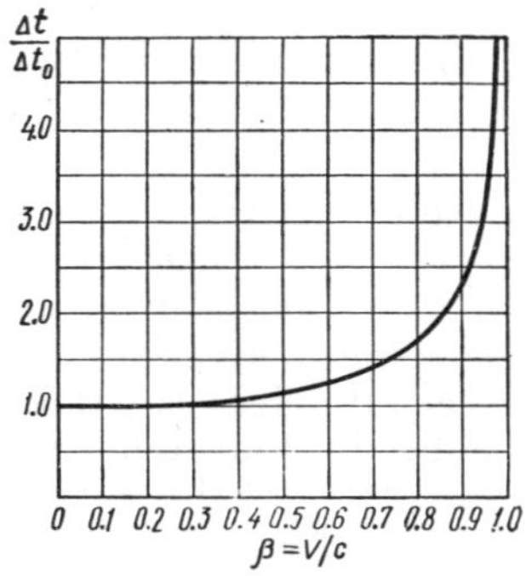

Fig. 111

Thus, as distinct from classical mechanics the rate of time flow actually depends on the state of motion. There is no such thing as universal time, and the notion of the "time interval between two given events" proves to be relative. The statement that a certain number of seconds has passed between two given events is meaningful only when the reference frame, relative to which it is valid, is indicated.

The absolute time of classical mechanics is an approximate notion in the theory of relativity holding only for low (compared to the velocity of light) relative velocities of reference frames. This follows directly from Eq. (6.4) and is seen from Fig. 111: $\Delta t = \Delta t_0$ at $V \ll c$.

So, we reach a fundamental conclusion: in a reference frame moving together with a clock, time flows slower (from the standpoint of the observer with respect to whom the given clock moves). The same is true for all processes proceeding in reference frames moving with respect to an observer.

Naturally, one may ask whether an observer in the K' frame, which moves relative to the K frame, realizes that his clock goes slower compared to the clock of the K frame. Obviously, he does not. This immediately follows from the principle of relativity. If the K' observer also noticed the dilation of time in his reference frame, that would mean that for *both* observers K' and K time flows more slowly in one of the inertial reference frames. From that fact the observers might have inferred that one of the inertial reference frames differed from another, which is in contradiction with the principle of relativity.

From this it follows that the dilation of time is reciprocal and symmetric relative to both inertial reference frames K and K'. In other words, if in terms of the K frame the clock of the K' frame goes slower, then in terms of the K' frame it is the clock of the K frame that goes slower (and with the same deceleration factor). This circumstance is evidence that the *dilation of time is a purely kinematic phenomenon*. It is an obligatory consequence of the velocity of light being invariant and cannot be attributed to some variation of clock properties caused by motion.

Eq. (6.4) has been experimentally confirmed by explaining the seemingly mysterious behaviour of muons traversing the Earth's atmosphere. Muons are unstable elementary particles whose average lifetime is $2 \cdot 10^{-6}$ s, this time being measured when the particles are at rest or moving with low velocities. Muons are generated in the upper layers of the atmosphere at a height of 20 to 30 km. Were the lifetime of muons independent of velocity, they would travel, even moving with the velocity of light, only about $c\Delta t = 3 \cdot 10^8 \cdot 2 \cdot 10^{-6}$ m $= 600$ m. Observations indicate, however, that a substantial number of muons nevertheless reach the Earth's surface.

The explanation is that the interval $2 \cdot 10^{-6}$ s is the proper lifetime Δt_0 of muons, time measured by a clock moving together with them. Since the velocity of these particles approaches that of light a time interval measured by a clock on Earth is much longer (see Eq. (6.4)) and turns out to be sufficient for muons to reach the Earth's surface.

In conclusion a few words about the so-called "*clock paradox*", or "*twin paradox*". Suppose there are two identical clocks A and B, with clock A resting in some inertial reference frame and clock B first moving away from clock A and then returning to it. The clocks are assumed to show the same time at the original moment, when they are located side by side.

In terms of clock A clock B moves, and therefore its rate is slower and during its travel it will lag behind clock A. But in terms of clock B it is clock A that moves and therefore on its return it will turn out to be slow. This evident incongruity constitutes the substance of the "paradox".

Actually, we made a fundamental mistake when we argued in terms of clock B since the reference frame fixed to that clock is non-inertial (it first moves away with acceleration and then comes back), and we cannot in this case use results related only to inertial reference frames. A detailed calculation lying outside the special theory of relativity indicates that the clock moving with acceleration (in our case clock B) goes slower, and therefore it is this clock that is behind when it returns to its initial point.

Lorentz contraction. Suppose rod AB moves relative to the K reference frame with the constant velocity V (Fig. 112) and the length of the rod in the reference frame K' fixed to the rod is equal to l_0. Our task is to determine the length l of that rod in the K frame.

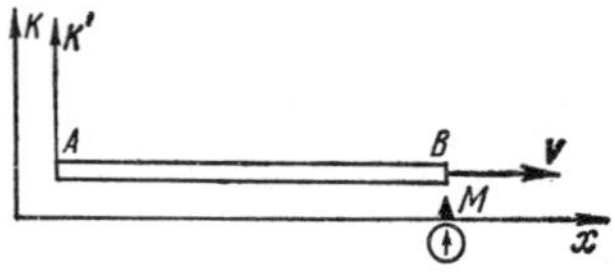

Fig. 112

For this purpose let us conduct the following imaginary experiment. Let us mark a point M on the x axis of the K frame and place a clock at that point. Using that clock we can measure the time of flight Δt_0 of the rod past the point M. Then we may state the sought length of the rod in the K frame to be equal to

$$l = V\Delta t_0.$$

An observer fixed to the rod registers a different time of flight. In fact, from his point of view the clock which registered the time of flight Δt_0 moves with the velocity

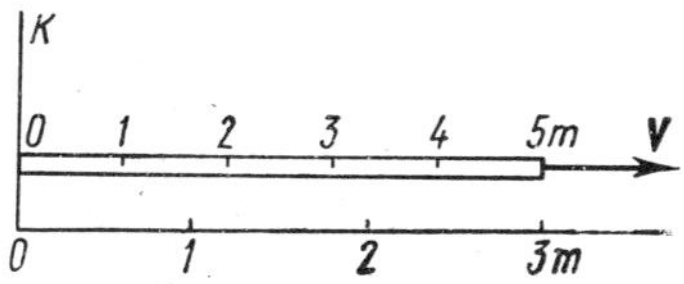

Fig. 113

V and therefore registers "someone else's" time. In accordance with Eq. (6.4) the observer's "own" time of flight is longer:

$$l_0 = V\Delta t.$$

From these two equations and Eq. (6.4) we get

$$l/l_0 = \Delta t_0/\Delta t = \sqrt{1-\beta^2}$$

or

1291

$$\boxed{l = l_0\sqrt{1-\beta^2},} \qquad (6.5)$$

where $\beta = V/c$. The length l_0 measured in the reference frame where the rod is at rest is referred to as the *proper length.*

Thus, the longitudinal length of a moving rod turns out to be shorter than its proper length, i.e. $l < l_0$. This phenomenon is called the *Lorentz contraction.* Note that this contraction occurs only in longitudinal dimensions of bodies, that is, dimensions along the motion direction, whereas the transverse dimensions do not vary, as it was shown above. As compared to the shape of a body in the reference frame where it is at rest, its shape in the moving reference frame may be characterized as oblate in the motion direction.

The length contraction of a moving rod is illustrated in Fig. 113, in which it is seen that in the reference frame fixed to the rod its length $l_0 = 5$ m, whereas in the K frame relative to which the rod moves with the velocity $V = 4/5\ c$ its length $l = 3$ m.

It follows from Eq. (6.5) that the degree of contraction depends on the velocity V. This dependence becomes especially pronounced at velocities V comparable with the velocity of light (Fig. 114).

So, in different inertial reference frames the length of the same rod turns out to be different. In other words, *length is a relative notion* which is meaningful only with respect to one or another reference frame. A statement about the length of a body being equal to some number of metres has no sense unless the reference frame relative to which the length is measured is indicated.

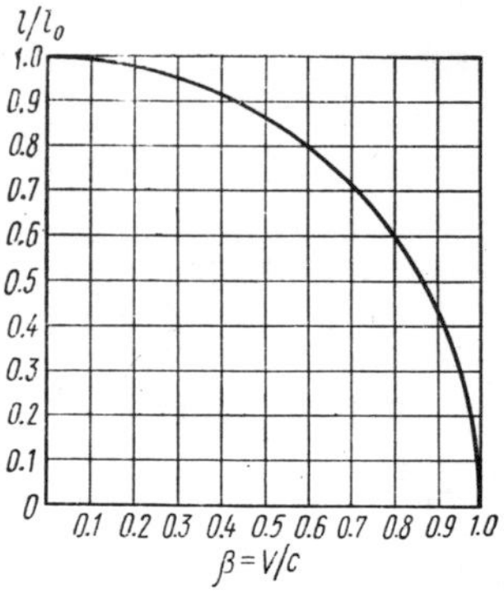

Fig. 114

As it follows from Eq. (6.5) and is seen in Fig. 114 $l \approx l_0$ at low velocities ($V \ll c$), so that the length of a body acquires an almost absolute meaning.

It should be pointed out that the Lorentz contraction, just as the dilation of time, must be reciprocal. This means that if we compare two rods moving relative to each other and having equal proper lengths, in terms of each of the rods the length of the other one is shorter in the same proportion. Were it not so, there would have appeared an experimental possibility to distinguish between the inertial reference frames fixed to the rods, which, however, contradicts the principle of relativity.

This points to the fact that the *Lorentz contraction is also a purely kinematic phenomenon*: no stresses causing deformations appear in a body.

It should be emphasized that the Lorentz contraction of bodies in the direction of their motion as well as the dilation of time are real and objective facts by no means associated with any illusions of an observer. All the values of dimensions of a given body and of time intervals obtained in different reference frames are equivalent, that is, all of them are "true". These statements are difficult to comprehend only because of our habit, based on routine experience, to regard length and time intervals as absolute notions while actually this is not so. The notions of length and time interval are as relative as those of motion and rest.

§ 6.4. Lorentz Transformation

Now we have to solve the fundamental problem of transformation of coordinates and time. Here we mean the formulae relating the coordinates and time moments of the same event in different reference frames.

Could the Galilean transformation possibly serve the purpose? Recall that this transformation is based on the assumption that the length of bodies is invariable and time flows at the same rate in different reference frames. In the previous section, however, we found out that in fact this is not so: the rate of time flow and the length of bodies depend on the reference frame. These are inevitable consequences of Einstein's postulates. Therefore, we are compelled to reject the Galilean transformation, or, more precisely, to recognize it as a special case of some more general transformation.

Thus, we have to obtain the transformation formulae which (i) take into account the dilation of time and the Lorentz contraction (i.e. are, in the final analysis, consequences of Einstein's postulates), and (ii) reduce to the Galilean transformation formulae in the limiting case of low velocities. Let us proceed to the solution of this problem.

We shall consider two inertial reference frames K and K'. Suppose the K' frame moves with the velocity $\mathbf{V}$ relative to the K frame. Let us orient the coordinate axes of the two frames as shown in Fig. 115: the x and x' axes coincide and are directed in parallel with the vector $\mathbf{V}$, and the y and y' axes are parallel to each other. Let us position at various points of the two reference frames identical clocks and synchronize them, separately the clocks of the K frame and the clocks of the K' frame. And finally, let us adopt the moment when the origins O and O' coincide as the zero time reading in both frames ($t = t' = 0$).

Suppose now that at the moment t at the point with the coordinates x, y in the K frame a certain event A takes place, e.g. a bulb flashes. Our task is to determine the coordinates x', y' and the time moment t' of that event in the K' frame.

The problem concerning the y' coordinate was solved at the beginning of this section, where it was shown (see Eq. (6.3)) that $y' = y$. Therefore, we immediately pass to searching the x' coordinate of the event. The x' coordinate describes the proper length of the segment $O'P$ resting in the K' frame (Fig. 115). The length of the same segment in the K frame where the measurement is taken at the moment t is equal to $x - Vt$. The relationship between these lengths is specified by Eq. (6.5), from which it follows that $x - Vt = x'\sqrt{1 - \beta^2}$. Whence

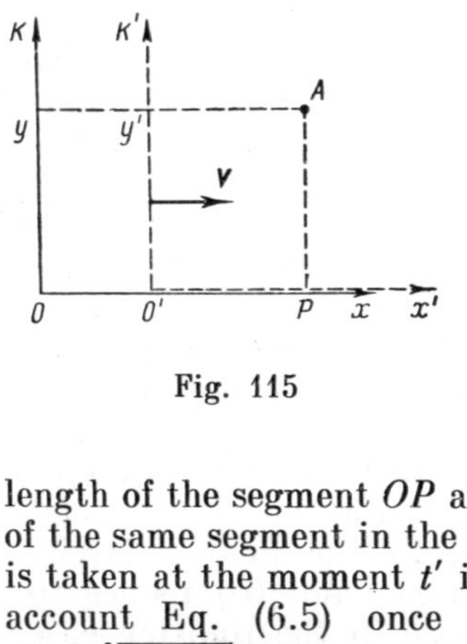

Fig. 115

$$x' = (x - Vt)/\sqrt{1 - \beta^2}. \quad (6.6)$$

On the other hand, the x coordinate describes the proper length of the segment OP at rest in the K frame. The length of the same segment in the K' frame where the measurement is taken at the moment t' is equal to $x' + Vt'$. Taking into account Eq. (6.5) once again, we obtain $x' + Vt' = x\sqrt{1 - \beta^2}$, whence

$$x = (x' + Vt')/\sqrt{1 - \beta^2}. \quad (6.6')$$

The obtained formulae make it possible to determine the relationship between the time moments t and t' of the event A in both reference frames. For this purpose it is sufficient to eliminate x' or x from Eqs. (6.6) and (6.6'), whereupon we get:

$$t' = (t - xV/c^2)/\sqrt{1 - \beta^2}; \quad t = (t' + x'V/c^2)/\sqrt{1 - \beta^2}. \quad (6.7)$$

Eqs. (6.3), (6.6), (6.6') and (6.7) are referred to as the *Lorentz transformation*. They play a key part in the theory of relativity. Using these formulae, the coordinates and time of any event can be transformed on transition from one inertial reference frame to another.

Thus, the Lorentz transformation on transition from the K frame to the K' frame takes the form

$$\boxed{x' = \frac{x - Vt}{\sqrt{1-\beta^2}}; \quad y' = y; \quad t' = \frac{t - xV/c^2}{\sqrt{1-\beta^2}},} \tag{6.8}$$

and for the reverse transition from the K' frame to the K frame

$$\boxed{x = \frac{x' + Vt'}{\sqrt{1-\beta^2}}; \quad y = y'; \quad t = \frac{t' + x'V/c^2}{\sqrt{1-\beta^2}},} \tag{6.9}$$

where $\beta = V/c$ and V is the velocity of the K' frame relative to the K frame.

It should be immediately emphasized that the symmetry (a similar form) of Eqs. (6.8) and (6.9) is the consequence of the complete equivalence of both reference frames (the different sign of V in these formulae is only due to the opposite motion direction of the K and K' frames relative to each other).

The Lorentz transformation differs drastically from the Galilean transformation (6.1), but the latter can be obtained from Eqs. (6.8) and (6.9) if the formal substitution $c = \infty$ is made in them. What does this mean?

At the end of the foregoing section it was mentioned that the Galilean transformation is based on the assumption of clocks synchronized by means of signals propagating instantaneously. From this fact it follows that the quantity c plays in the Lorentz transformation the part of the velocity of the signals utilized for clock synchronization. When this velocity is infinitely great, we get the Galilean transformation; when it is equal to the velocity of light, the Lorentz transformation. Thus, the Lorentz transformation is based on the assumption of clock synchronization by means of light signals possessing ultimate velocity.

The remarkable feature of the Lorentz transformation

is the fact that at $V \ll c$ it reduces* to the Galilean transformation (6.1). Thus, in the limiting case $V \ll c$ the transformation laws of the theory of relativity and classical mechanics coincide. This means that the theory of relativity does not reject the Galilean transformation as incorrect but includes it into the true transformation laws as a special case that is valid at $V \ll c$. In what follows we shall see that this reflects the general relationship between the theory of relativity and classical mechanics: the laws and relations of the theory of relativity turn into the laws of classical mechanics in the limiting case of low velocities.

Next, from the Lorentz transformation it is seen that at $V > c$ the radicands become negative and the formulae lose physical meaning. This corresponds to the fact that bodies cannot move with a velocity exceeding that of light *in vacuo*. It is even impossible to use a reference frame moving with the velocity $V = c$; in this case the radicands turn into zero and the formulae lose physical meaning. This means that no reference frame can in principle be fixed, for example, to a photon moving with the velocity c. Expressed otherwise, there is no reference frame in which a photon is at rest.

And finally, it should be noted that the time transformation formulae contain a spatial coordinate. This significant circumstance reveals the inseparable relationship between space and time. In other words, we should not speak separately of space and time but of unified *space-time* in which all physical phenomena take place.

§ 6.5. Consequences of Lorentz Transformation

Concept of simultaneity. Suppose two events A_1 (x_1, y_1, t_1) and A_2 (x_2, y_2, t_2) occur in the reference frame K. Let us find the time interval separating these events in the K' frame moving with the velocity V along the x axis as

* Strictly speaking, it is also required that $x/c \ll t$, i.e. the times of propagation of light signals over the distances typical for the problems considered (x/c) should be less than the time intervals we discuss here. When this condition is satisfied, the signals may be regarded as propagating instantaneously.

shown in Fig. 115. In accordance with the time transformation formula (6.8) the sought time interval is equal to

$$t_2' - t_1' = \frac{(t_2 - t_1) - (x_2 - x_1)\, V/c^2}{\sqrt{1-\beta^2}}. \tag{6.10}$$

From this it follows that events which are simultaneous in the K frame ($t_2 = t_1$) are not simultaneous in the K' frame ($t_2' - t_1' \neq 0$). The only exception is the case when the two events occur in the K frame simultaneously at points with the same values of the x coordinate (the y coordinate may have any value).

Simultaneity is thus a relative concept; events which are simultaneous in one reference frame are not simultaneous in the general case in another reference frame. When discussing simultaneity of events, one has to specify the reference frame relative to which simultaneity occurs. Otherwise, the concept of simultaneity loses its meaning.

It follows from relativity of simultaneity that clocks positioned along the x' axis in the K' frame and synchronized together in that reference frame show different time in the K frame. Indeed, for the sake of simplicity let us consider the moment when the origins O and O' of the two reference frames coincide and the clocks at those points show the same time: $t = t' = 0$. Then in the K frame the clock at a point with the coordinate x shows at that moment the time $t = 0$, while the clock of the K' frame at the same point shows a different time, t'. Indeed, in accordance with the time transformation formula (6.8)

$$t' = -xV/c^2 \sqrt{1-\beta^2}.$$

It is seen from this that at the moment $t = 0$ (in the K frame) the clock of the K' frame shows a different time depending on the x coordinate (the so-called *local time*). This is shown in Fig. 116*a*. In terms of the K' frame the situation is reciprocal (Fig. 116*b*), as it in fact should be due to the equivalence of both inertial reference frames.

Next, it is seen from Eq. (6.10) that for events simultaneous in the K frame the sign of the difference $t_2' - t_1'$ is defined by the sign of the expression $-(x_2 - x_1)\, V$. Consequently, in different reference frames (with different values of the velocity V) the difference $t_2' - t_1'$ is different not only in magnitude but also in sign. This fact signifies that the sequence of the events A_1 and A_2 may be either direct or reverse.

This, however, does not apply to events obeying the causality principle. The sequence of such events (cause → → effect) is the same throughout all reference frames. This can be easily demonstrated through the following reasoning. Let us consider, for example, a firing, event A_1 (x_1, t_1), and a hitting of a target with a bullet, event A_2 (x_2 , assuming that both events occur on the x axis. In the K frame $t_2 > t_1$, the velocity of the bullet is v, and assuming,

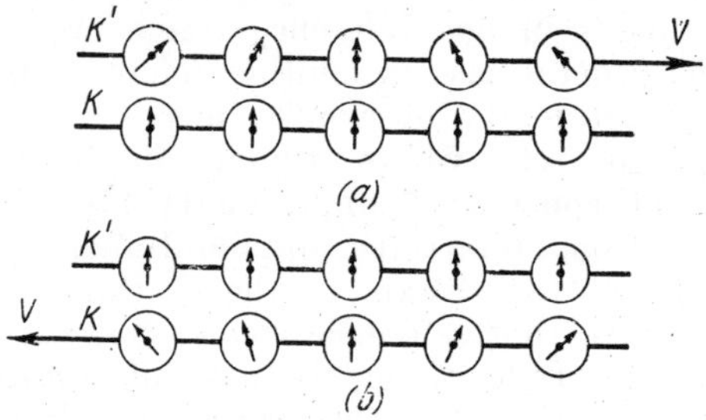

Fig. 116

for definiteness, $x_2 > x_1$, we may write $x_2 - x_1 = v(t_2 - t_1)$. Substituting this equality into Eq. (6.10), we get

$$t'_2 - t'_1 = (t_2 - t_1)(1 - vV/c^2)/\sqrt{1-\beta^2}.$$

The quantity in the second parentheses of the numerator is always positive as $V < c$ (even at $v = c$ when the cause-and-effect relationship is determined by signals or interactions propagating with the highest possible velocity). It follows that if $t_2 > t_1$, then $t'_2 > t'_1$, i.e. the sequence of the cause-and-effect events is the same in all inertial reference frames.

Lorentz contraction. Let us orient a rod resting in the K' frame along the x' axis, i.e. along the motion direction of this reference frame relative to the K frame. Suppose the length of the rod in the K' frame is equal to $l_0 = x'_2 - x'_1$ (the proper length).

In the K frame, relative to which the rod moves, its length is *defined* as the distance l between the coordinates x_2 and x_1 of its ends taken *at the same moment of time* ($t_2 = t_1$). Making use of the Lorentz transformation (6.8) for the x' and x coordinates, we get

$$l_0 = x_2' - x_1' = (x_2 - x_1)/\sqrt{1-\beta^2} = l/\sqrt{1-\beta^2},$$

whence

$$l = l_0 \sqrt{1-\beta^2}. \tag{6.11}$$

Thus, the length l of the moving rod proves to be less than its proper length l_0, and in each reference frame it has its own value. This result is in a complete agreement with the result obtained in Eq. (6.5).

It follows from the definition of the length that the relativity of the length of the given rod is a consequence of the relativity of simultaneity. The same pertains to the form of any body: its dimensions in the motion direction are also different in different inertial reference frames.

Duration of processes. Suppose that at the point with the coordinate x' of the reference frame K' a certain process takes place whose duration in that frame is equal to $\Delta t_0 = t_2' - t_1'$ (the proper time of the process). Let us determine the duration of the given process $\Delta t = t_2 - t_1$ in the K frame relative to which the K' frame moves.

For this purpose we shall resort to the Lorentz transformation of time. As the process takes place at the point with the fixed coordinate x' of the K' frame, it is more convenient to use Eq. (6.9):

$$t_2 - t_1 = (t_2' - t_1')/\sqrt{1-\beta^2}$$

or

$$\Delta t = \Delta t_0/\sqrt{1-\beta^2}. \tag{6.12}$$

It is seen that the duration of the same process is different in different inertial reference frames. In the K frame its duration is longer ($\Delta t > \Delta t_0$), and therefore in that reference frame it proceeds slower than in the K' frame. This fact is in a complete agreement with the result concerning the rate of the same clock in different inertial reference frames, that is, Eq. (6.4).

Interval. The relativity of spatial and time intervals by no means implies that the theory of relativity denies the existence of any absolute quantities. In fact, the opposite is true. The theory of relativity tackles the problem of finding such quantities (and laws) which are independent of the choice of the inertial reference frame.

The first of these quantities is the universal velocity with which interactions propagate; it is equal to the velocity of light *in vacuo*. The second invariant value, just as important, is the so-called *interval* s_{12} between events *1* and *2*, whose square is defined as

$$\boxed{s_{12}^2 = c^2 t_{12}^2 - l_{12}^2 = \text{inv},} \qquad (6.13)$$

where t_{12} is the time interval between the events and l_{12} is the distance between the two points at which the given events occur ($l_{12}^2 = x_{12}^2 + y_{12}^2 + z_{12}^2$).

We can easily see that the interval is invariant, calculating it directly in the reference frames K and K'. Making use of the Lorentz transformation (6.8) and taking into account that $y'_{12} = y_{12}$ and $z'_{12} = z_{12}$, we can write:

$$c^2 t'^2_{12} - x'^2_{12} = c^2 \frac{(t_{12} - x_{12} V/c^2)^2}{1-\beta^2} - \frac{(x_{12} - V t_{12})^2}{1-\beta^2} = c^2 t_{12}^2 - x_{12}^2.$$

Thus, it is clear that the interval is really invariant. In other words, the statement "two events are separated by a certain interval s" has an absolute meaning for it is valid in all inertial reference frames. The invariant interval plays a fundamental role in the theory of relativity and provides an efficient instrument of analysis and solution of many problems (see, e.g., Problem 6.4).

Types of intervals. Depending on what component, spatial or temporal, prevails, an interval is referred to as either

space-like ($l_{12} > ct_{12}$), or

time-like ($ct_{12} > l_{12}$).

In addition to these two types of intervals there is another type, light-like ($ct_{12} = l_{12}$).

If the interval between two events is space-like, a reference frame K' can always be found in which these events occur simultaneously ($t'_{12} = 0$):

$$c^2 t_{12}^2 - l_{12}^2 = -l'^2_{12}.$$

If the interval is time-like, one can always find a reference frame K' in which both events occur at the same point ($l'_{12} = 0$):

$$c^2t_{12}^2 - l_{12}^2 = c^2t_{12}'^2.$$

In the case of space-like intervals $l_{12} > ct_{12}$, i.e. the events cannot influence each other in any reference frame, even if communication between the events is carried out at the ultimate velocity c. This is not the case with time-like and light-like intervals, for which $l_{12} \leqslant ct_{12}$. Consequently, events separated by such intervals may be in a cause-and-effect relationship with each other.

Transformation of velocity. Suppose in the x, y plane of the K frame a particle moves with the velocity **v**, whose projections are equal to v_x and v_y. Using the Lorentz transformation (6.8), we find the velocity projections v'_x and v'_y of that particle in the K' frame moving with the velocity **V** as shown in Fig. 115:

$$v'_x = \frac{dx'}{dt'} = \frac{dx'}{dt}\,\frac{1}{dt'/dt} = \frac{v_x - V}{1 - v_xV/c^2},$$

$$v'_y = \frac{dy'}{dt'} = \frac{dy'}{dt}\,\frac{1}{dt'/dt} = \frac{v_y\sqrt{1-\beta^2}}{1 - v_xVc^2}, \tag{6.14}$$

where $\beta = V/c$. Hence, the velocity of the particle in the K' frame is

$$v' = \sqrt{v_x'^2 + v_y'^2} = \frac{\sqrt{(v_x - V)^2 + v_y^2(1-\beta^2)}}{1 - v_xV/c^2}. \tag{6.15}$$

These formulae yield the so-called *relativistic law of transformation of velocity*. At low velocities ($V \ll c$ and $v \ll c$) they reduce, as one can easily see, to the classical formulae of transformation of velocity:

$$v'_x = v_x - V; \quad v'_y = v_y,$$

or in a vector form

$$\mathbf{v}' = \mathbf{v} - \mathbf{V}.$$

Note that the last formula is valid only in the Newtonian approximation; in the relativistic case it has no meaning, for the simple law of velocity composition cannot be applied here. This can easily be demonstrated by the following example. Suppose the velocity vector **v** of a particle in the K frame is perpendicular to the x axis, i.e. has the projections $v_x = 0$ and $v_y = v$. Then in accordance with Eq. (6.14) the velocity projections of the same particle in the K' frame are

$$v'_x = -V; \quad v'_y = v_y \sqrt{1-\beta^2}. \tag{6.16}$$

This means that in the given case ($\mathbf{v} \perp x$ axis) the v'_y projection diminishes on transition to the K' frame and clearly $\mathbf{v}' \neq \mathbf{v} - \mathbf{V}$ (Fig. 117).

Let us now examine the case when two particles move toward each other with the same velocity v in the K reference frame. What is the velocity v' of one particle with respect to the other? In the non-relativistic approximation the obvious answer is $2v$. In the case of high velocities we have to apply the first formula of (6.14), assuming the particles to move along the x axis. Let us fix the K' reference frame to one of the particles, e.g. to the one moving in the positive direction of the x axis. Then the problem reduces to finding the velocity of the other particle in that reference frame. Substituting $v_x = -v$ and $V = v$ into Eq. (6.14), we obtain

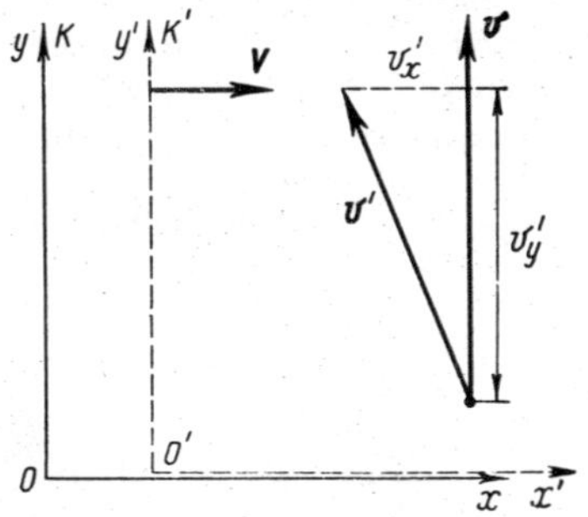

Fig. 117

$$v'_x = -2v/[1+(v/c)^2].$$

The minus sign means that in the given case the second particle moves in the negative direction of the x' axis of the K' reference frame. It should be pointed out that even when both particles move almost with the highest possible velocity $v \cong c$, the velocity v' cannot exceed c (which is immediately seen from the last formula).

And finally, let us check directly whether the relativistic formulae of velocity transformation correspond to Einstein's second postulate concerning the constancy of the velocity of light c in all inertial reference frames. Suppose vector c has the projections c_x and c_y in the K frame, i.e. $c^2 = c_x^2 + c_y^2$. Now let us transform the radicand in Eq. (6.15) as follows:

$$c_x^2 - 2c_xV + V^2 + (c^2 - c_x^2)\left(1 - \frac{V^2}{c^2}\right) = \left(c - \frac{c_xV}{c}\right)^2.$$

After this it is not difficult to get $v' = c$. Of course, in the general case vector $\mathbf{c}'$ is oriented differently in the K' frame.

§ 6.6. Geometric Description of Lorentz Transformation

Let us now consider the relativistic concepts of space-time using the geometric approach developed by Minkowski, which helps to describe the substance of the Lorentz transformation in a different light.

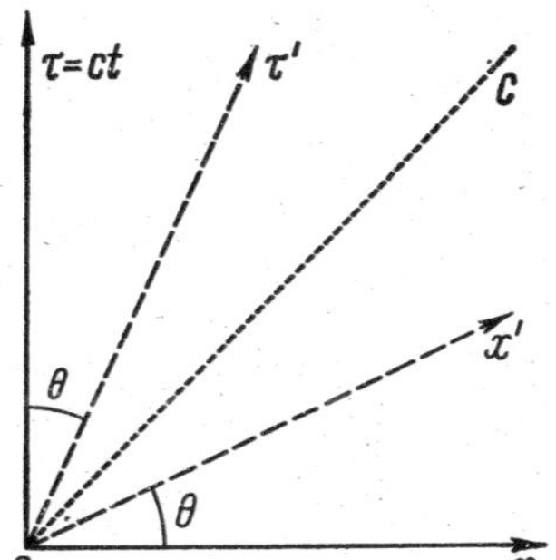

Fig. 118

Minkowski diagrams. Suppose there are two inertial reference frames K and K', the latter moving relative to the former with the velocity V. First, let us draw the so-called space-time diagram for the K frame, confining ourselves to the more simple and graphic unidimensional case (Fig. 118). In this diagram the ordinate axis usually marks not the time t itself but the quantity $\tau = ct$ (where c is the velocity of light). We may thus calibrate both axes, Ox and $O\tau$, in metres using the same scale.

Each point of the diagram, referred to as a *world point*, describes a certain event A (x, τ). Each particle (even a stationary one) has a corresponding *world line* in this diagram. For example, the $O\tau$ axis is the world line of a particle resting at the point $x = 0$. The Ox axis depicts the totality of events simultaneous with the event O irrespective of the x coordinate.

The world line corresponding to light propagating from the point O in the positive direction of the x axis is the bisector OC of the right angle (the dotted line in Fig. 118).

Let us plot the τ' and x' axes of the K' frame in this diagram. Assuming $x' = 0$ in the Lorentz transformation (6.8), we obtain the world line of the origin of the K' frame. Then $x = Vt = \beta\tau$, where $\beta = V/c$. This is the equation of the straight line forming the angle θ with the τ axis, which can be determined from the formula $\tan\theta = \beta$. The straight line obtained is the world line depicting the totality of events occurring at the origin of the K' frame, i.e. it is the τ' axis.

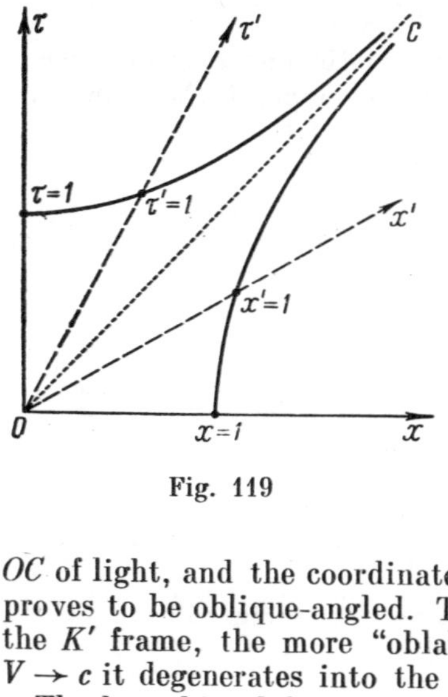

Fig. 119

The x' axis of the K' frame is a straight line depicting all the events which are simultaneous with the event O in the K' frame. Assuming $t' = 0$ in the Lorentz transformation (6.8), we get $ct = xV/c$, or $\tau = \beta x$. From this it follows that the x' axis forms the same angle θ ($\tan\theta = \beta$) with the x axis.

Thus, the τ' and x' axes of the K' frame are arranged symmetrically with respect to the world line OC of light, and the coordinate grid (τ', x') of the K' frame proves to be oblique-angled. The higher the velocity V of the K' frame, the more "oblate" its coordinate grid is. As $V \to c$ it degenerates into the world line of light.

The last thing left to be done in the diagram is to scale the τ, x, τ', x' axes of both reference frames. The easiest way to do this is to utilize the invariance of the interval:

$$s^2 = \tau^2 - x^2 = \tau'^2 - x'^2.$$

Let us mark on the τ axis of the K frame the point corresponding to a time unit in the K frame ($\tau = 1$, Fig. 119) and then draw through that point the hyperbola

$$\tau^2 - x^2 = 1,$$

whose points conform to the invariant interval $s = 1$ (since when $x = 0$, $\tau = 1$ and $s = 1$). Its asymptote is the world line of light. The hyperbola crosses the τ' axis at the point corresponding to a time unit in the K' frame. Indeed, $\tau'^2 - x'^2 = 1$ and if $x' = 0$, then $\tau' = 1$.

The x and x' axes are calibrated in much the same way: a hyperbola $x^2 - \tau^2 = 1$ is drawn through the point $x = 1$, $\tau = 0$ of the K frame; then the point at which the hyperbola crosses the x' axis, and where $\tau' = 0$, marks a unit of length ($x' = 1$) in the K' frame (since $x'^2 - \tau'^2 = 1$ and if $\tau' = 0$, then $x' = 1$).

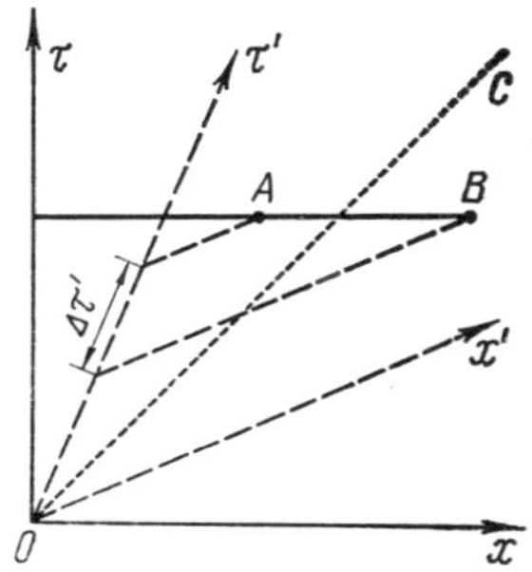

Fig. 120

The Minkowski diagram thus plotted illustrates the transition from the K to K' frame and conforms to the Lorentz transformation (6.8). In accordance with the principle of relativity the reverse transition from the K' to K frame is illustrated by a diagram of a quite symmetric form: the coordinate grid of the K' frame is rectangular and the K frame is oblique-angled. We suggest that the readers themselves should demonstrate this.

Now we shall show how the Minkowski diagram assists in interpreting simply and graphically such relativistic effects as, for example, the relativity of simultaneity, the dilation of time, and the Lorentz contraction.

Relativity of simultaneity follows immediately from Fig. 120. In fact, events A and B simultaneous in the K frame turn out to occur at different moments in the K' frame. The event A occurs later than event B by the time $\Delta\tau'$.

Dilation of time. Let us consider two clocks K and K' which show the same time $\tau = \tau' = 0$ at the moment when they are at the same point in space ($x = x' = 0$). The clock K is assumed to be at rest in the K frame and the clock K' in the K' frame.

Suppose that according to the clock K a time unit elapses ($\tau = 1$); this corresponds to the event A in the diagram

(Fig. 121). Let us draw a hyperbola $\tau^2 - x^2 = 1$ through the point A and also a straight line AB' describing all the events which are simultaneous with the event A in the K frame. The intersection of the τ' axis, i.e. the world line of the clock K', with the hyperbola gives the point A' ($\tau' = 1$) and with the straight line OB' the point B' ($\tau' < 1$). This means that in the K' frame a time unit has not yet

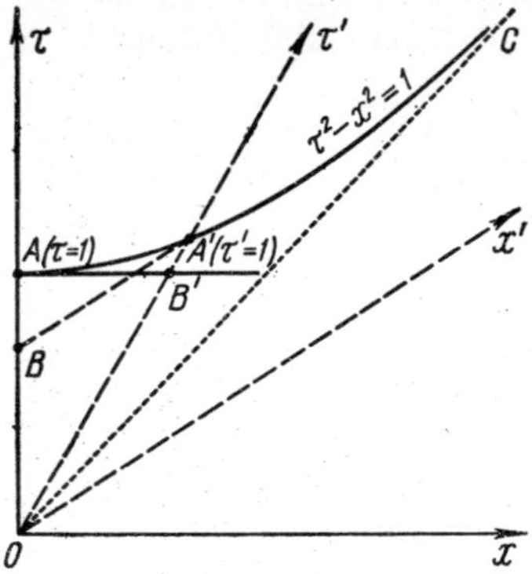

Fig. 121

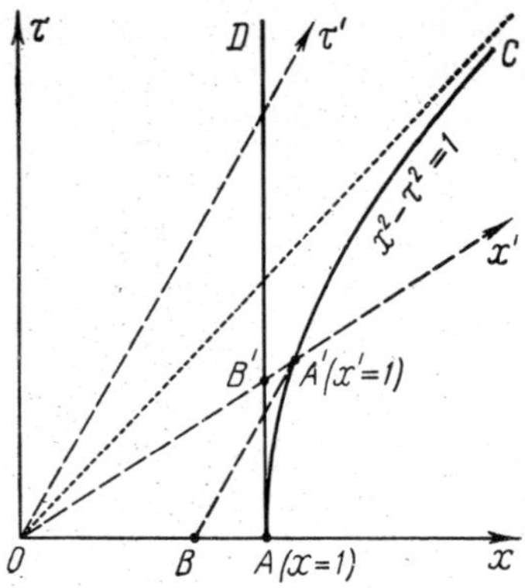

Fig. 122

elapsed according to the moving clock K' when the clock K registers the passing of a time unit. This fact signifies that the clock K' goes slower.

Let us utilize the diagram to make sure that the effect of the dilation of time is reversible. Draw a straight line BA' parallel to the x' axis which describes all the events which are simultaneous with the event A' in the K' frame ($\tau' = 1$). The point B at which this straight line intersects the world line of the clock K, the τ axis, shows that $\tau < 1$, i.e. it is the clock K whose rate is now slower with respect to the K' frame.

Lorentz contraction. Suppose a one-metre rod is at rest in the K frame (the segment OA in Fig. 122). The straight lines $O\tau$ and AD are the world lines of its ends. To measure the length of that rod in the K' frame, we have to determine the coordinates of its ends simultaneously in that frame.

But in the K' frame the event O (the determination of the left end of the rod) is simultaneous with the event B' represented by the point of intersection of the world line of the right end of the rod with the simultaneity line Ox'. It is seen from the diagram that in the K' frame $OB' < OA'$, i.e. the rod moving relative to the K' frame is shorter than one metre.

It can be demonstrated just as easily that the Lorentz contraction is reversible as well. If a one-metre rod is at rest in the K' frame (segment OA'), then, drawing the world lines of its ends in that frame ($O\tau'$ and $A'B$), we see that in the K frame $OB < OA$ provided the coordinates of the rod's ends are determined simultaneously, i.e. the K' rod experiences the Lorentz contraction with respect to the K frame.

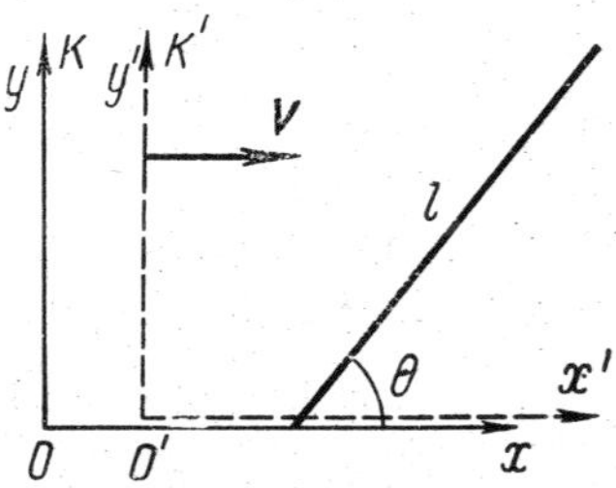

Fig. 123

Problems to Chapter 6

6.1. A stationary rod of length $l = 1.00$ m is oriented at the angle $\theta = 45°$ to the x axis of the K frame (Fig. 123). Find its length l' and the corresponding angle θ' in the K' frame moving relative to the K frame with the velocity $V = c/2$ along the x axis.

Solution. The rod's length in the K' frame is

$$l' = \sqrt{(\Delta x')^2 + (\Delta y')^2} = \sqrt{(\Delta x)^2 (1-\beta^2) + (\Delta y)^2}.$$

Taking into account that $\Delta x = l \cos\theta$ and $\Delta y = l \sin\theta$, we get

$$l' = l\sqrt{1-\beta^2 \cos^2\theta} = 0.94 \text{ m}.$$

The angle θ' in the K' frame is found from its tangent:

$$\tan\theta' = \frac{\Delta y'}{\Delta x'} = \frac{\Delta y}{\Delta x\sqrt{1-\beta^2}} = \frac{\tan\theta}{\sqrt{1-\beta^2}}; \qquad \theta' = 49°.$$

It should be pointed out that the results obtained are independent of the direction of the velocity of the K' frame.

6.2. A rod moves along a ruler with a certain constant velocity. When the positions of both ends of the rod are determined simultaneously in the reference frame fixed to the ruler, the length of the rod $l_1 = 4.0$ m. However, when the positions of the ends of the rod are determined simultaneously in the reference frame fixed to the rod, the difference of readings made by the ruler is equal to $l_2 = 9.0$ m. Find:

(1) the proper length of the rod;

(2) the velocity of the rod relative to the ruler.

Solution. The proper length of the rod l_0 is related to l_1 and l_2 via the following formulae:

$$l_1 = l_0 \sqrt{1-\beta^2}; \quad l_0 = l_2 \sqrt{1-\beta^2},$$

where β is the velocity of the rod expressed in units of the velocity of light. From these formulae we obtain:

(1) $l_0 = \sqrt{l_1 l_2} = 6$ m; (2) $\beta = \sqrt{1 - l_1/l_2} = \sqrt{5}/3 \approx 0.75$ or $v = 0.75\ c$.

● **6.3. Rod-and-tube "paradox".** A tube AB of length 1.0 m is at rest in the K frame. Let us take a rod $A'B'$ of length 2.0 m and accelerate it to such a velocity that its length in the K frame becomes equal to 1.0 m. Then at a certain moment the rod, flying through the

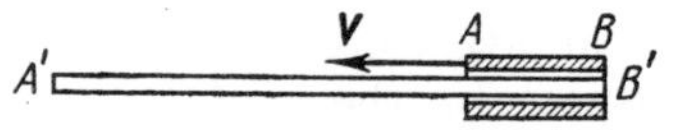

Fig. 124

tube, fits in it completely. However, "in terms of" the rod it is the tube that becomes reduced by half, and consequently the rod (2.0 m) does not fit in the tube (0.5 m). Is there a contradiction here?

Solution. "In terms of" the tube the ends of the flying rod coincide with the ends of the tube simultaneously. "In terms of" the rod the ends do not coincide simultaneously: first, the ends B and B' coincide (Fig. 124), and after the time interval Δt, the ends A and A'. The time interval Δt may be calculated as follows:

$$\Delta t = (L_0 - l)/V = 6 \cdot 10^{-9} \text{ s},$$

where $L_0 = 2.0$ m is the proper length of the rod, $l = 0.5$ m is the length of the tube moving relative to the rod, and V is its velocity. The latter is found from Eq. (6.11): $V = c\sqrt{3}/2$.

● **6.4.** Find the distance which an unstable particle traverses in the K frame from the moment of its generation till decay, provided its lifetime in that reference frame is $\Delta t = 3.0 \cdot 10^{-6}$ s and its proper lifetime is $\Delta t_0 = 2.2 \cdot 10^{-6}$ s.

Solution. Using Eq. (6.12), we determine the velocity V of the particle and then the distance sought:

$$l = \Delta t \cdot V = \Delta t \cdot c\sqrt{1-(\Delta t_0/\Delta t)^2} = 0.6 \text{ km}.$$

Another method of solution is based on the invariance of the interval:

$$c^2 (\Delta t_0)^2 = c^2 (\Delta t)^2 - l^2,$$

where on the left-hand side we write the squared interval in the frame fixed to the particle and on the right-hand side the squared interval in the K frame. From this we obtain the same value of l.

● **6.5. Doppler effect.** A stationary detector P of light signals is located in the K frame (Fig. 125). A source S of light signals approaches the detector with the velocity V. In the reference frame fixed to the source the signals are emitted with the frequency ν_0. With what frequency ν does the detector receive these signals?

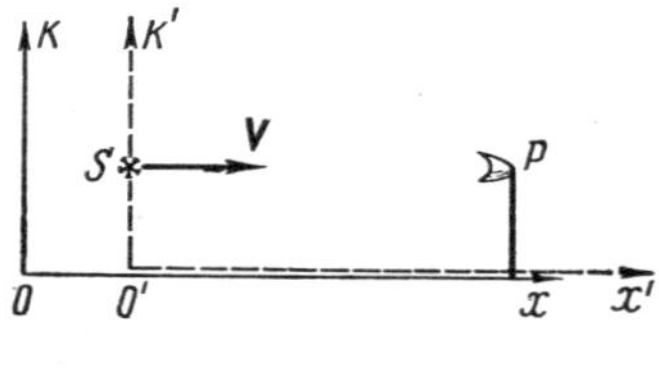

Fig. 125

Solution. The time interval between two consecutive signals (pulses) in the K' frame fixed to the source is equal to $T_0 = 1/\nu_0$. As this frame moves with the velocity V, the corresponding time interval in the K frame is, in accordance with Eq. (6.12), longer:

$$T = T_0/\sqrt{1-\beta^2}, \quad \beta = V/c$$

The distance between two consecutive pulses in the K frame is equal to

$$\lambda = cT - VT = (c-V)\,T = (c-V)\frac{T_0}{\sqrt{1-\beta^2}}. \quad (1)$$

Therefore, the frequency received by the detector is equal to

$$\nu = \frac{c}{\lambda} = \frac{c\sqrt{1-\beta^2}}{T_0(c-V)} = \nu_0\sqrt{\frac{1+\beta}{1-\beta}}$$

or

$$\nu = \nu_0\frac{\sqrt{1-\beta^2}}{1-\beta}.$$

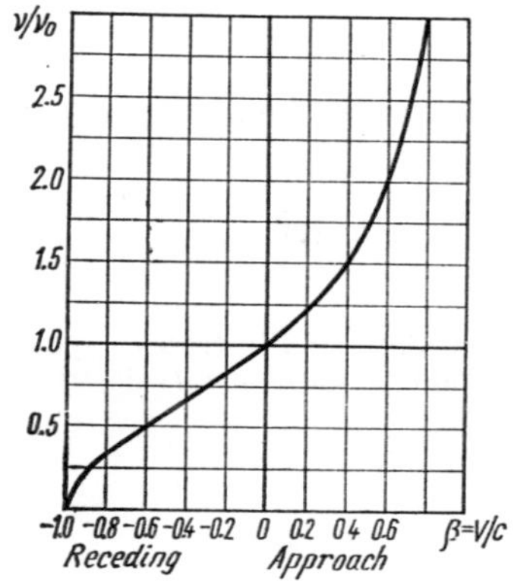

Fig. 126

When the source approaches (as in our case), then $\nu > \nu_0$, but when it moves away, $\nu < \nu_0$ (in this case the sign of β changes to the opposite, Fig. 126). The formula obtained for the frequency ν expresses the so-called *radial Doppler effect*.

Note that in classical physics $T = T_0$ since time is absolute. Therefore, the classical formula for the Doppler effect does not contain the factor $\sqrt{1-\beta^2}$, which is replaced by unity:

$$\nu = \nu_0/(1-\beta) \approx \nu_0(1 + V/c).$$

At the same time let us consider a more general case: in the K frame the velocity V of the source forms the angle α with the line of observation (Fig. 127). In this case it is sufficient to replace V in

Eq. (1) by $V \cos\alpha$. Then

$$\nu = \nu_0 \frac{\sqrt{1-\beta^2}}{1-\beta\cos\alpha}.$$

In particular, at $\alpha = \pi/2$ the so-called *transverse Doppler effect* is observed:

$$\nu = \nu_0 \sqrt{1-\beta^2},$$

in which the observed frequency always proves to be lower than the proper one ν_0. Incidentally, the last expression is just a consequence

Fig. 127

of the dilation of time in a moving reference frame; it may also be obtained directly from Eq. (6.12):

$$\nu = \frac{1}{T} = \frac{1}{T_0/\sqrt{1-\beta^2}} = \nu_0 \sqrt{1-\beta^2}.$$

● **6.6. Relationships between events.** Fig. 128 illustrates a *space-time diagram*. Each point of that diagram (a *world point*) describes

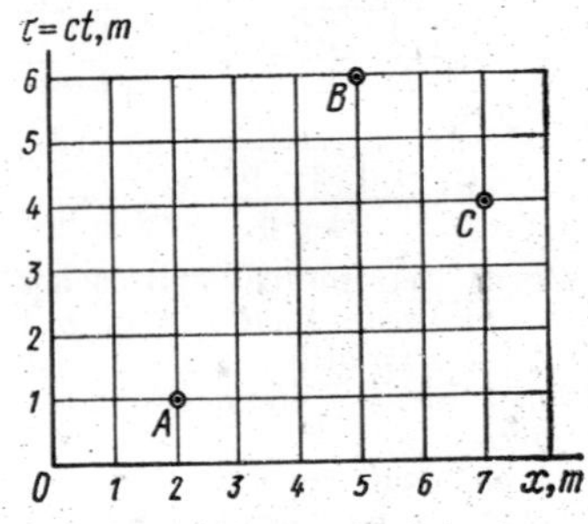

Fig. 128

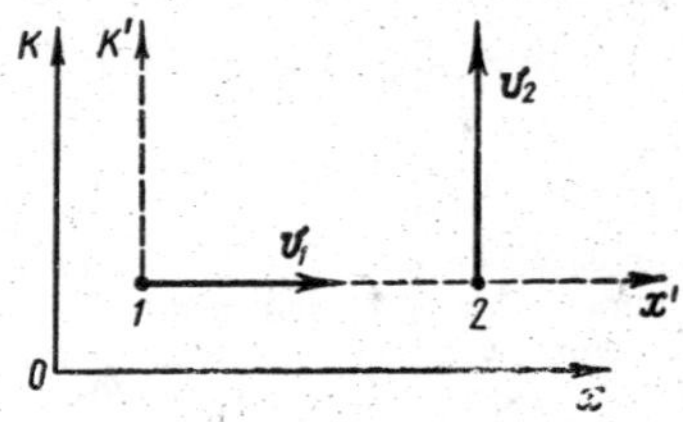

Fig. 129

a certain event, that is, a coordinate and a time moment at which that event happens. Let us examine three events corresponding to the world points A, B, and C. Prove that the following relationships exist between these events:

Pair of events	Type of interval	Proper time $c \cdot \Delta t_0$, m	Proper distance Δx_0, m	Possible cause-and-effect relation
AB	Time-like	4	—	$A \to B$
AC	Space-like	—	4	None
BC	Light-like	0	0	$C \to B$

Hint: use invariance of the interval.

6.7. Two particles move in the K frame at right angles to each other: the first one with the velocity v_1 and the second one with v_2. Find the velocity of one particle with respect to the other.

Solution. Let us choose the coordinate axes of the K frame as illustrated in Fig. 129. When the K' frame is fixed to particle *1*, the velocity of particle *2* in that reference frame is the value sought. Introducing $V = v_1$ and $v_x = 0$ in Eq. (6.15), we get

$$v_2' = \sqrt{v_{2x}'^2 + v_{2y}'^2} = \sqrt{v_1^2 + v_2^2 - (v_1 v_2/c)^2}.$$

Note that in accordance with the classical law of vector composition

$$v_2' = \sqrt{v_1^2 + v_2^2}.$$

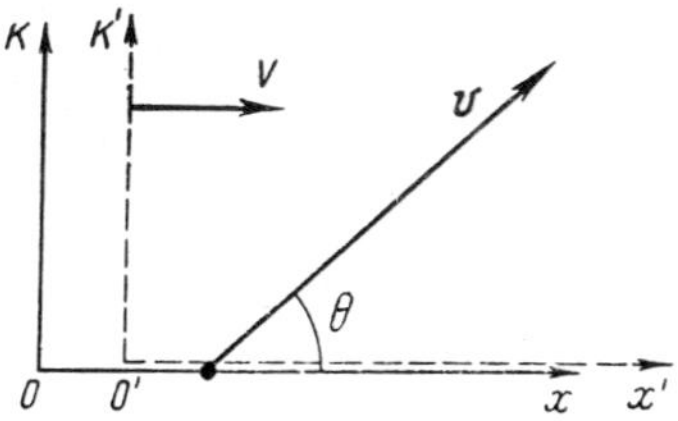

Fig. 130

6.8. Velocity direction transformation. A particle moves in the K frame with the velocity $\mathbf{v}$ at the angle θ to the x axis. Find the corresponding angle θ' in the K' frame, which moves with the velocity $\mathbf{V}$ as illustrated in Fig. 130.

Solution. Suppose the projections of the vector $\mathbf{v}$ in the K frame are equal to v_x and v_y. Then the following relation is true:

$$\tan \theta' = v_y / v_x.$$

Taking into account Eqs. (6.14) we obtain in the K' frame

$$\tan \theta' = v_y'/v_x' = v_y \sqrt{1-\beta^2}/(v_x - V).$$

After the substitution $v_x = v \cos \theta$ and $v_y = v \sin \theta$ we find

$$\tan \theta' = \frac{\sin \theta \cdot \sqrt{1-\beta^2}}{\cos \theta - V/v}.$$

As the last equation shows, the angle transformation law for velocity differs from that for segments (see Problem 6.1).

6.9. A rod oriented parallel to the x axis of the K reference frame moves in that frame with the velocity v in the positive direction of the y axis. Find the angle θ' between the rod and the x' axis of the K' frame travelling with the velocity V relative to the K frame in the positive direction of its x axis. The x and x' axes coincide, the y and y' axes are parallel to each other.

Solution. Suppose that at a certain moment the ends of the rod coincide with the x axis in the K frame. These two events, which are simultaneous in the K frame, are not so in the K' frame; in accordance with Eq. (6.10) they are separated by the time interval

$$\Delta t' = \Delta x V/c^2\sqrt{1-\beta^2},$$

where Δx is the proper length of the rod. In that time the right end of the rod shifts "higher" than the left one by $\Delta y' = v_y' \Delta t'$, where $v_y' = v\sqrt{1-\beta^2}$ (see Eq. (6.16)). Thus, in the K' frame the rod is turned counterclockwise through the angle θ', which may be determined using the formula

$$\tan\theta' = \Delta y'/\Delta x' = \beta v/c\sqrt{1-\beta^2},$$

where $\Delta x' = \Delta x\sqrt{1-\beta^2}$ is the projection of the rod on the x' axis of the K' frame, and $\beta = V/c$.

● **6.10. Relativistic transformation of acceleration.** A particle moves with the velocity **v** and the acceleration **w** in the K frame. Find the acceleration of that particle in the K' frame, which shifts with the velocity **V** in the positive direction of the x axis of the K frame. Examine the cases when the particle moves along the following axes of the K frame: (1) x, (2) y.

Solution. 1. Let us write each projection of the acceleration of the particle in the K' frame as follows:

$$w_x' = \frac{dv_x'}{dt'} = \frac{dv_x'}{dt}\cdot\frac{1}{dt'/dt}.$$

Making use of the first of formulae (6.14) and the last one of (6.8), we get after differentiation:

$$w_x' = \frac{(1-\beta^2)^{3/2}}{(1-\beta v_x/c)^3}\,w_x; \qquad w_y' = 0.$$

2. Similar calculations yield the following results:

$$w_x' = 0; \qquad w_y' = (1-\beta^2)\,w_y.$$

In these formulae $\beta = V/c$.

CHAPTER 7

RELATIVISTIC DYNAMICS

§ 7.1. Relativistic Momentum

Let us first recall two basic assumptions of Newtonian mechanics concerning momentum:

(1) the momentum of a particle is defined as $\mathbf{p} = m\mathbf{v}$, and the mass m of the particle is supposed to be independent of its velocity;

(2) the momentum of a closed system of particles does not vary with time in any inertial reference frame.

Now we shall turn to relativistic dynamics. It is found here that the conservation law for Newtonian momentum is not valid in the case of a closed system of relativistic particles. (We shall illustrate this later by a simple example.) Thus we face the following choice: either to reject the Newtonian definition of momentum, or to discard the law of conservation of that quantity.

Considering the immense significance of the conservation laws, in the theory of relativity the momentum conservation law is regarded as fundamental and the momentum itself is expressed accordingly*.

First of all we shall demonstrate that the requirement for the momentum conservation law to hold in any inertial reference frame together with the relativistic transformation of velocities on transition from one inertial reference frame to another leads to the conclusion that the particle's mass must depend on the velocity of the particle (in contrast

* The following question arises: how can the momentum conservation law be of any value if momentum is defined so as to keep it constant? To answer the question, let us imagine a particle colliding with other particles in the process of its motion. Having considered the first collision, we *define* momentum so that it obeys the conservation law in that collision. In the following collisions, however, the situation is different: we know the momenta of the particles involved in those collisions, and the momentum conservation law (if it really exists) is valid not according to definition but due to underlying laws of nature.

Experience shows that momentum *thus* defined really obeys the conservation law. At least, not a single phenomenon has been observed up to now in which that law fails.

to Newtonian mechanics). For this purpose let us examine a completely inelastic collision of two particles, where the system is assumed closed.

Suppose two *identical* particles *1* and *2* move toward each other in an inertial reference frame K with the same velocity v_0 at an angle α to the x axis (Fig. 131*a*). In that reference frame the total momentum of both particles apparently remains constant: it equals zero both before and after the collision (and the formed particle turns out to be motionless, as follows from symmetry considerations).

Now let us see what happens in another inertial reference frame. First, we choose two reference frames: the K_1 frame

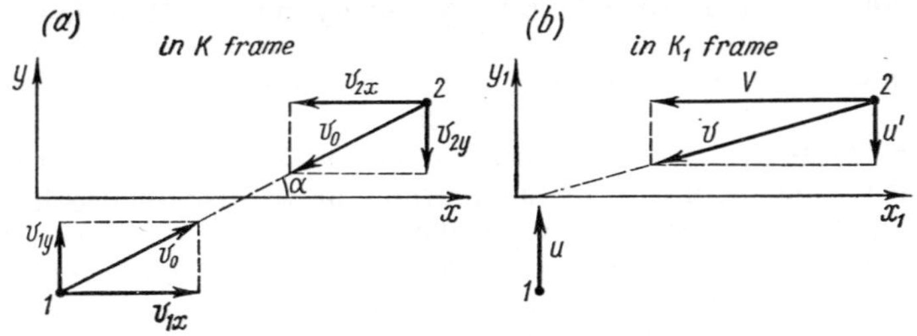

Fig. 131

moving to the right with the velocity v_{1x} and the K_2 frame moving to the left with the velocity v_{2x} (Fig. 131*a*). Clearly, particle *1* in the K_1 frame and particle *2* in the K_2 frame move only along the y axis with velocities whose equal moduli we denote by u.

Let us consider the collision in the K_1 frame (Fig. 131*b*) in which particle *1* has the velocity u. We find the y component of the velocity of particle *2* in that reference frame, denoting it by u'. As we mentioned, that particle moves with the velocity u along the y axis in the K_2 frame and at the same time translates together with the K_2 frame to the left with the velocity V relative to the K_1 frame. Therefore, in accordance with Eq. (6.16), the y component of the velocity of particle *2* in the K_1 frame is equal to

$$u' = u\sqrt{1-(V/c)^2}. \tag{7.1}$$

Now the y components of momenta of both particles may be written in the K_1 frame as m_1u and m_2u'. In accordance with Eq. (7.1) $u' < u$, and therefore it is easy to see that the momentum conservation law does not hold true in its conventional (Newtonian) statement. Indeed, in our case $m_1 = m_2$ (the particles being identical) and therefore the y component of the total momentum of the particles before the collision differs from zero while after the collision it is equal to zero, since the particle formed moves only along the x axis.

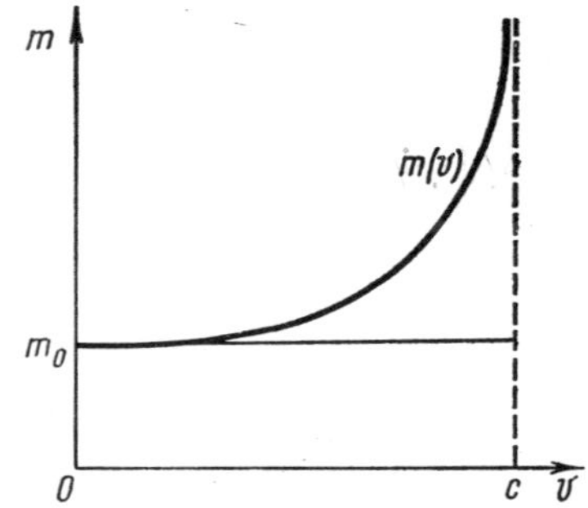

Fig. 132

The momentum conservation law becomes valid in the K_1 frame if we assume $m_1u = m_2u'$. Then from Eq. (7.1) we get

$$m_2 = m_1/\sqrt{1-(V/c)^2}.$$

When $\alpha \to 0$ (Fig. 131), $u \to 0$ and m_1 is the mass of the motionless particle; it is denoted by m_0 and is referred to as the *rest mass*. In that case the velocity V proves to be equal to v, the velocity of particle *2* with respect to particle *1*. Consequently, the last formula can be rewritten as

$$m = m_0/\sqrt{1-(v/c)^2}, \tag{7.2}$$

where m is the mass of the moving particle (recall that the two particles are identical). The mass m is referred to as *relativistic*. As it is seen from Eq. (7.2), the latter is greater than the rest mass and depends on the particle's velocity (Fig. 132).

Thus, we have reached an important conclusion: *the relativistic mass of a particle depends on its velocity*. In other words, the mass of the same particle is different in different inertial reference frames.

In contrast to the relativistic mass the particle's rest mass m_0 is an invariant quantity, i.e. is the same in all reference frames. For this reason we can claim that rest mass is a characteristic property of a particle. But later on we shall frequently use the relativistic mass m to simplify some conclusions, reasonings, and calculations.

Now we shall take the last step. Using Eq. (7.2) we write the momentum of a relativistic particle in the following form:

$$\mathbf{p} = m\mathbf{v} = \frac{m_0\mathbf{v}}{\sqrt{1-(v/c)^2}}. \qquad (7.3)$$

This is the so-called *relativistic momentum* of a particle. Experience confirms that the momentum *thus* defined does obey the conservation law regardless of the inertial reference frame chosen.

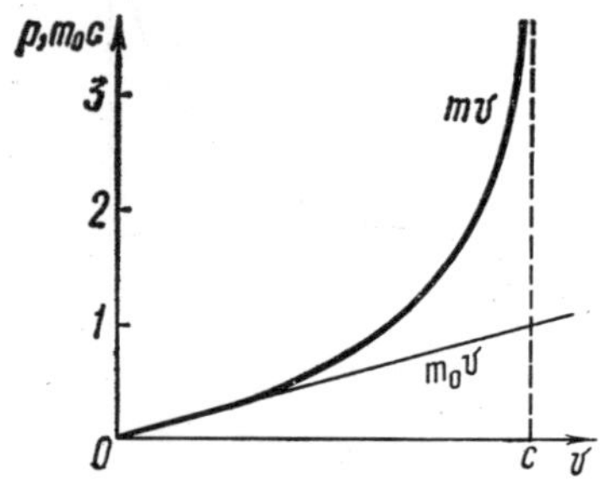

Fig. 133

Note that if $v \ll c$ Eq. (7.3) yields the Newtonian definition of momentum: $\mathbf{p} = m_0\mathbf{v}$, where m_0 is independent of the velocity v. The velocity dependences of the relativistic and the Newtonian momentum of a particle are compared in Fig. 133. The difference between the momenta is seen to grow substantially as the velocity of a particle approaches that of light.

Let us consider two examples illustrating how Eqs. (7.2) and (7.3) are applied.

Example 1. In modern giant accelerators protons can be accelerated up to a velocity differing from that of light by 0.01 per cent. How many times does the relativistic mass of such protons exceed their rest mass?

In accordance with Eq. (7.2) $m/m_0 = 1/\sqrt{1-\beta^2}$, where $\beta = v/c$. Since β slightly differs from unity, the radicand should be transformed as follows:

$$1-\beta^2 = (1+\beta)(1-\beta) \approx 2(1-\beta).$$

Then

$$m/m_0 \approx 1/\sqrt{2(1-\beta)} \approx 70.$$

Example 2. At what velocity does a particle's Newtonian momentum differ from its relativistic one by one per cent? by ten per cent?

From the condition $\eta = (p - p_{cl})/p = 1 - \sqrt{1 - (v/c)^2}$ we get

$$v/c = \sqrt{\eta\,(2 - \eta)}.$$

Whence

$$\frac{v}{c} = \begin{cases} 0.14 \text{ at } \eta = 0.01, \\ 0.45 \text{ at } \eta = 0.10. \end{cases}$$

Thus, the classical formula for momentum provides an accuracy better than one per cent at $v/c \leqslant 0.14$ and better than ten per cent at $v/c \leqslant 0.45$.

§ 7.2. Fundamental Equation of Relativistic Dynamics

According to Einstein's principle of relativity all laws of nature must be invariant with respect to inertial reference frames. In other words, the mathematical formulations of laws must be identical in all these reference frames. In particular, this is true for the laws of dynamics.

However, detailed analysis shows that the fundamental equation of dynamics of Newton $m\mathbf{w} = \mathbf{F}$ does not satisfy Einstein's principle of relativity. The Lorentz transformation totally changes the form of the equation on transition to another inertial frame.

To satisfy the requirements of the principle of relativity, the fundamental equation of dynamics must have another form and only in the case of $v \ll c$ turn into the Newtonian equation. It is shown in the theory of relativity that these requirements are met by the equation

$$d\mathbf{p}/dt = \mathbf{F}, \tag{7.4}$$

where $\mathbf{F}$ is the force acting on the particle. This equation completely coincides in *form* with the fundamental equation of Newtonian dynamics (4.1). But the physical meaning is different here: the left-hand side of the equation contains the time derivative of a *relativistic* momentum defined by formula (7.3). Substituting Eq. (7.3) into Eq. (7.4), we obtain

$$\boxed{\frac{d}{dt}\left(\frac{m_0\mathbf{v}}{\sqrt{1-(v/c)^2}}\right) = \mathbf{F}.} \tag{7.5}$$

This is the *fundamental equation of relativistic dynamics.*

It can be easily seen that the equation written in this form ensures the invariance of momentum for a free particle and turns into the fundamental equation of Newtonian dynamics ($m\mathbf{w} = \mathbf{F}$) at low velocities ($v \ll c$).

Moreover, the fundamental equation of dynamics, when written in this form, proves to be invariant relative to the Lorentz transformation and, consequently, satisfies Einstein's principle of relativity. We shall not prove this here, but shall only note that on transition from one inertial reference frame to another the force **F** is transformed in accordance with definite rules. In other words, the force **F** is not an invariant in the theory of relativity and its magnitude and direction vary*.

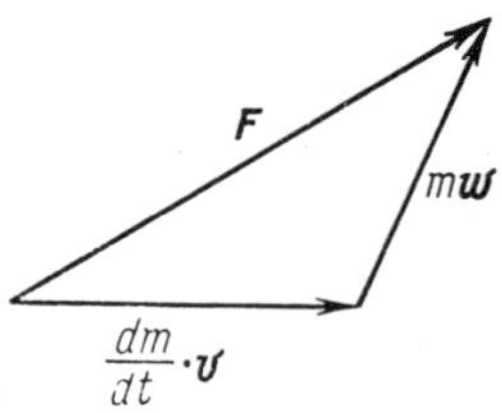

Fig. 134

A surprising conclusion follows from the fundamental equation of relativistic dynamics: the acceleration vector **w** of a particle does not coincide in the general case with the direction of the force vector **F**. To demonstrate this, we write Eq. (7.5) in the following form:

$$d(m\mathbf{v})/dt = \mathbf{F},$$

where m is the relativistic mass of the particle. Differentiating with respect to time, we obtain

$$(dm/dt)\,\mathbf{v} + m\,(d\mathbf{v}/dt) = \mathbf{F}. \tag{7.6}$$

This expression is graphically illustrated in Fig. 134. Thus, the acceleration vector **w** is indeed not collinear with the force vector **F** in the general case.

The acceleration **w** coincides in direction with the vector **F** only in two cases:

(1) $\mathbf{F} \perp \mathbf{v}$ (*transverse force*); in this case the magnitude of the vector **v** does not vary, i.e. $v = \text{const}$, and Eq. (7.5)

* As distinct from Newtonian mechanics where forces are absolute, in the theory of relativity the force projections perpendicular to the direction of the relative velocity vector of the reference frames are different in these frames. The projections have maximum values in the reference frame where the particle is at rest at a given moment:

$$F'_x = F_x, \quad F'_y = F_y\sqrt{1-(v/c)^2}.$$

takes the form

$$m_0\mathbf{w}/\sqrt{1-(v/c)^2}=\mathbf{F},$$

whence the acceleration

$$\mathbf{w}=(\mathbf{F}/m_0)\sqrt{1-(v/c)^2};$$

(2) $\mathbf{F}\parallel\mathbf{v}$ (*longitudinal force*). In this case Eq. (7.5) may be written in scalar form; performing differentiation with respect to time on the left-hand side of the equation, we obtain

$$\left(\frac{m_0}{\sqrt{1-(v/c)^2}}+\frac{m_0v^2/c^2}{[1-(v/c)^2]^{3/2}}\right)\frac{dv}{dt}=F,$$

whence the acceleration written in vector form is

$$\mathbf{w}=(\mathbf{F}/m_0)[1-(v/c)^2]^{3/2}.$$

It is not difficult to see that if the force F and the velocity v have the same values in both cases, the transverse force imparts to the particle a greater acceleration than the longitudinal force.

The fundamental equation of relativistic dynamics makes it possible to find the law relating to the force $\mathbf{F}$ acting on a particle provided the time dependence of the relativistic momentum $\mathbf{p}(t)$ is known, and on the other hand, to find the equation of motion of the particle $\mathbf{r}(t)$ if the acting force and the initial conditions, the velocity $\mathbf{v}_0$ and the position $\mathbf{r}_0$ at the initial moment of time, are known.

The application of Eq. (7.5) is illustrated by problems 7.1-7.3.

§ 7.3. Mass-Energy Relation

Kinetic energy of a relativistic particle. We shall define this quantity in the same fashion as we did in classical mechanics, i.e. as a quantity whose increment is equal to the work performed by the force acting on a particle. First we find the increment of the particle's kinetic energy dT due to the force $\mathbf{F}$ acting over the elementary path $d\mathbf{r} = \mathbf{v}dt$:

$$dT = \mathbf{F}\mathbf{v}dt.$$

In accordance with the fundamental equation of relativistic dynamics (7.4) $\mathbf{F}dt = d(m\mathbf{v}) = dm \cdot \mathbf{v} + md\mathbf{v}$, where m is the relativistic mass. Therefore,

$$dT = \mathbf{v}(dm \cdot \mathbf{v} + md\mathbf{v}) = v^2 dm + mvdv,$$

where the relation $\mathbf{v}d\mathbf{v} = vdv$ is taken into account (see p. 90). This expression can be simplified by allowing for the dependence of mass on velocity (Eq. (7.2)). Squaring the equation, we get

$$m^2c^2 = m^2v^2 + m_0^2c^2.$$

Let us find the differential of this expression, bearing in mind that m_0 and c are constants:

$$2mc^2dm = 2mv^2dm + 2m^2vdv.$$

The right-hand side of this equality, when divided by $2m$, coincides with the expression for dT. Hence, it follows that

$$dT = c^2dm. \tag{7.7}$$

Thus, the increment of the kinetic energy of a particle is proportional to the increment of its relativistic mass. The kinetic energy of a motionless particle is equal to zero and its mass is equal to the rest mass m_0. Consequently, integrating Eq. (7.7), we obtain

$$T = (m - m_0)c^2, \tag{7.8}$$

or

$$\boxed{T = m_0c^2\left(\frac{1}{\sqrt{1-\beta^2}} - 1\right),} \tag{7.9}$$

where $\beta = v/c$. This is the expression for the *relativistic kinetic energy* of a particle. It can be seen how conspicuously it differs from the classical $m_0v^2/2$. Let us make sure, however, that at low velocities ($\beta \ll 1$) expression (7.9) turns into the classical one. For this purpose we employ the binomial theorem, according to which

$$\frac{1}{\sqrt{1-\beta^2}} = (1-\beta^2)^{-1/2} = 1 + \frac{1}{2}\beta^2 + \frac{3}{8}\beta^4 + \ldots .$$

At $\beta \ll 1$ we can confine ourselves to the first two terms of the series, and then

$$T = m_0c^2\beta^2/2 = m_0v^2/2.$$

Thus, at high velocities the kinetic energy of a particle is given by the relativistic formula (7.9), which is different from $m_0v^2/2$. It should be pointed out here that expression (7.9) cannot be represented as $mv^2/2$, where m is the relativistic mass.

The relativistic T_{rel} and classical T_{cl} kinetic energies plotted as functions of β are compared in Fig. 135. Their difference becomes very pronounced at velocities comparable to that of light.

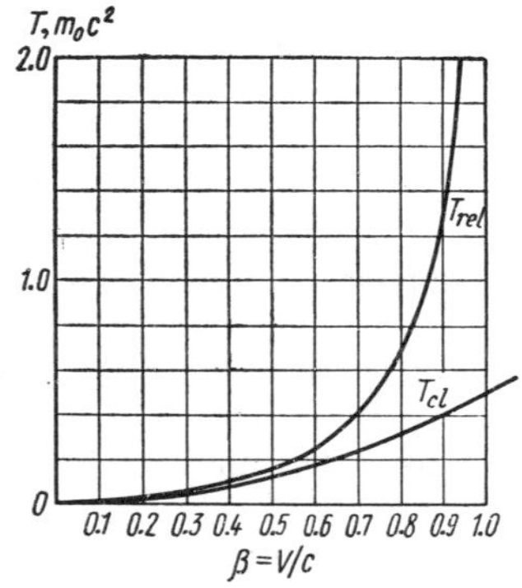

Fig. 135

Example 1. A particle of mass m_0 moves with the velocity at which its relativistic kinetic energy T exceeds by n times the kinetic energy calculated by means of the classical formula. Find T.

For the sake of simplicity we introduce the designation $\tau = T/m_0c^2$. Then the given condition $T = n \cdot m_0v^2/2$ may be written as

$$\tau = n\beta^2/2,$$

where $\beta = v/c$. From Eq. (7.9) β can be expressed as

$$\beta^2 = 1 - 1/(1 + \tau)^2.$$

Eliminating β^2 from these two equations, we get

$$2\tau^2 + (4 - n)\,\tau - 2\,(n - 1) = 0.$$

The root of this equation is

$$\tau = [n - 4 + \sqrt{n\,(n + 8)}]/4.$$

The minus sign in front of the radicand has no physical meaning (τ cannot be negative) and thus can be omitted.

Here are the four values of τ calculated from the last formula for the following n:

$n = T/T_{cl}$:	1.01	1.1	1.5	2.0
$\tau = T/m_0c^2$:	0.0067	0.065	0.32	0.62

It is seen that, for example, at $T/m_0c^2 \leqslant 0.0067$ the application of the classical formula permits the kinetic energy to be determined with an accuracy better than one per cent.

Example 2. What amount of work must be performed to increase the velocity of a particle of rest mass m_0 from 0.6 c to 0.8 c? Compare the result obtained with that calculated from the classical formula.

In accordance with Eq. (7.9) the work sought is equal to

$$A=T_2-T_1=m_0c^2\left(\frac{1}{\sqrt{1-\beta_2^2}}-\frac{1}{\sqrt{1-\beta_1^2}}\right)=0.42m_0c^2.$$

The classical formula yields the following value:

$$A=m_0(v_2^2-v_1^2)/2=0.14m_0c^2.$$

The difference between the two results is seen to be substantial.

Relation between mass and energy. It follows from Eq.(7.7) that the increment of kinetic energy of a particle is accompanied by a proportional increment of its relativistic mass. It is known, however, that various processes taking place in nature are connected with the transformation of one kind of energy into another. For example, the kinetic energy of colliding particles can be transformed into the internal energy of a new particle formed after the collision. Therefore, it is natural to expect that the mass of a body grows not only due to additional kinetic energy but also due to *any* increase in the total energy stored in the body, irrespective of what specific kind of energy is responsible for that increase.

Owing to this Einstein reached the following fundamental conclusion: the total energy of a body (or a system of bodies), whatever kinds of energy it comprises (kinetic, electric, chemical, etc.), is related to the mass of that body by the equation

$$\boxed{E=mc^2.} \tag{7.10}$$

This formula expresses one of the most fundamental laws of nature, the relationship (proportionality) of the mass m and the *total energy* E of a body. To avoid misunderstanding, we should point out that the total energy E does not include the potential energy of a body in an external field, provided such a field acts on the body.

Relation (7.10) may be written in another form if Eq. (7.8) is taken into account. Then the total energy of a body is

$$E = m_0c^2 + T,$$

where m_0 is the rest mass of a body and T is its kinetic energy. From this it follows directly that a motionless body ($T = 0$) also possesses the energy

$$E_0 = m_0c^2. \tag{7.11}$$

This energy is referred to as the *rest energy* or the *proper energy*.

We see that the mass of a body which in non-relativistic mechanics manifested itself as a measure of inertness (in Newton's second law), or as a measure of gravitational action (in the law of universal gravitation), emerges now as a measure of the *energy content* of a body. In accordance with the theory of relativity even a body at rest has a certain amount of stored-up energy, the rest energy.

A change in the total energy of a body (a system) is accompanied by an equivalent change in its mass $\Delta m = \Delta E/c^2$ and vice versa. In conventional macroscopic processes the change of mass of bodies turns out to be extremely small, so that its experimental detection is impossible. This can be demonstrated by the following examples.

Examples. A. A satellite of mass $m = 100$ kg is launched into the Earth's orbit by accelerating it to the velocity $v = 8$ km/s. This means that its energy increases by $\Delta E = mv^2/2$ (allowing for $v \ll c$). The corresponding increase in the satellite's mass is equal to

$$\Delta m = \Delta E/c^2 = mv^2/2c^2 = 3.5 \cdot 10^{-8} \text{ kg}.$$

B. Heating one litre of water from 0 to 100 °C requires the energy $\Delta E = mc_p\, \Delta t$, where $c_p = 4.2$ J/(g·K) is the specific heat of water and Δt is the temperature difference. The corresponding increase in the mass of the water is

$$\Delta m = \Delta E/c^2 = 0.47 \cdot 10^{-10} \text{ kg}.$$

C. A spring of stiffness factor $\varkappa = 10^3$ N/cm is compressed by $\Delta l = 1$ cm. In the process the spring acquires the energy $\Delta E = \varkappa\, (\Delta l)^2/2$. The equivalent increment in its mass is equal to

$$\Delta m = \Delta E/c^2 = 0.5 \cdot 10^{-16} \text{ kg}.$$

It is easy to see that in all three cases the mass changes lie far outside the capabilities of experimental technique.

In astronomical phenomena, however, associated, for example, with exploration of stars mass can change by an appreciable amount. One may ascertain this by the example of solar radiation.

Example. According to astronomical observations the amount of energy carried by solar radiation each second to an area of 1 m^2 of the Earth's surface oriented at right angles to the solar rays comes to about $1.4 \cdot 10^3$ J/(s·m^2). This makes it possible to calculate the total energy radiated by the Sun per second:

$$\Delta E = 1.4 \cdot 10^3 \cdot 4\pi R^2 = 4 \cdot 10^{26} \text{ J/s},$$

where R is the distance between the Earth and the Sun. Consequently every second the Sun loses the mass

$$\Delta m = \Delta E/c^2 = 4.4 \cdot 10^9 \text{ kg/s!}$$

This value is stupendous on the Earth's scale, but when compared to the mass of the Sun this loss is negligible: $\Delta m/m = 2 \cdot 10^{-21}$ s^{-1}.

Things are quite different in nuclear physics. Here it became possible for the first time to experimentally check and confirm the law relating mass and energy. This is because nuclear processes and transformations of elementary particles are associated with very large changes of energy comparable with the rest energy of the particles themselves. We shall return to this problem in § 7.5.

§ 7.4. Relation Between Energy and Momentum of a Particle

It is clear that both the energy E and the momentum p of a particle have different values in different reference frames. There is, however, a quantity, a certain combination of E and p, that is invariant, i.e. has the same value in different reference frames. Such a quantity is $E^2 - p^2c^2$. Let us make sure that this is so.

Making use of the formulae $E = mc^2$ and $p = mv$, we may write

$$E^2 - p^2c^2 = m^2c^4 - m^2v^2c^2 = \frac{m_0^2c^4}{1-(v/c)^2}[1-(v/c)^2]$$

or after cancelling

$$\boxed{E^2 - p^2c^2 = m_0^2c^4.} \tag{7.12}$$

The cancellation of the velocity v on the right-hand side means that the value of $E^2 - p^2c^2$ is independent of the velocity of the particle, and consequently, of the reference frame. In other words, the quantity $E^2 - p^2c^2$ is indeed an invariant with the value $m_0^2c^4$ in all inertial reference frames:

$$E^2 - p^2c^2 = \text{inv}. \tag{7.13}$$

This conclusion is of great importance since it allows, as it will be shown later, the analysis and solution of various problems to be drastically simplified in many cases.

Here are two more relations which are very often useful. The first one is

$$\boxed{\mathbf{p} = m\mathbf{v} = E\mathbf{v}/c^2} \tag{7.14}$$

and the second one relates the momentum and the kinetic energy T of a particle; it can be easily obtained by substituting $E = m_0c^2 + T$ into Eq. (7.12):

$$\boxed{pc = \sqrt{T(T + 2m_0c^2)}.} \tag{7.15}$$

At $T \ll m_0c^2$ the last relation turns into the classical one, $p = \sqrt{2m_0T}$, and when $T \gg m_0c^2$, it takes the form $p = T/c$.

Example. Assuming the rest energy of an electron to be equal to 0.51 MeV, calculate:

(1) the momentum* of an electron possessing a kinetic energy equal to its rest energy;

(2) the kinetic energy of an electron possessing the momentum 0.51 MeV/c, where c is the velocity of light.

1. If $T = m_0c^2$, we obtain from Eq. (7.15) $p = \sqrt{3}\, m_0c = 0.9$ MeV/c.

2. This problem may also be solved by resorting to Eq. (7.15). A simpler way, though, is to utilize Eq. (7.12):

$$T = E - m_0c^2 = \sqrt{p^2c^2 + m_0^2c^4} - m_0c^2 = 0.21 \text{ MeV}.$$

---

* Note that now momenta of relativistic particles are expressed in the units "energy/c", where c is the velocity of light. E.g., if energy is expressed in MeV units (1 MeV = $1.6 \cdot 10^{-6}$ erg), then momentum is in MeV/c. The introduction of such a unit for momentum simplifies many kinds of calculations quite noticeably.

In passing, let us examine the interesting possibility of the existence of particles with *zero rest mass* ($m_0 = 0$). From the equations

$$E = m_0c^2/\sqrt{1-(v/c)^2} \quad \text{and} \quad p = m_0v/\sqrt{1-(v/c)^2}$$

it follows that a particle whose rest mass $m_0 = 0$ may possess energy and momentum only when it moves with the velocity of light c. Then the last two formulae turn into the indeterminate ratio 0/0. This fact does not signify, however, the indeterminacy of energy and momentum of such a particle. The point is that both these quantities prove to be independent of velocity. Moreover, the relationship between the momentum p and the energy E is specified by Eq. (7.14), where $v = c$, i.e.

$$p = E/c. \tag{7.16}$$

Thus, in accordance with the theory of relativity the existence of particles with zero rest mass is possible, provided they move with the velocity c. This motion is not a result of preceding acceleration but this is the only state in which such particles can ever exist. The stoppage of such a particle is equivalent to its absorption (disappearance). At the present time there are two such particles known: the photon and the neutrino.

The Lorentz transformation of momentum and energy. Let a particle move with the velocity $v = dl/dt$ in the K reference frame. From Eq. (6.13) it follows that the elementary interval is

$$ds = \sqrt{c^2(dt)^2 - (dl)^2} = c\,dt\,\sqrt{1-(v/c)^2}.$$

Bearing that expression in mind, we present the projections of the momentum and the energy of the particle in the following form:

$$p_x = \frac{m_0}{\sqrt{1-(v/c)^2}}\,\frac{dx}{dt} = m_0c\,\frac{dx}{ds}; \qquad p_y = m_0c\,\frac{dy}{ds};$$

$$E = \frac{m_0c^2}{\sqrt{1-(v/c)^2}}\,\frac{dt}{dt} = m_0c^3\,\frac{dt}{ds} = m_0c\,\frac{c^2\,dt}{ds}.$$

From the invariance of the interval ds it immediately follows that on transition to another inertial reference frame p_x and p_y are transformed as dx and dy, i.e. as x and y, whereas the energy E is transformed as $c^2\,dt$, i.e. as the time t. Thus, the following correlations may be pointed out:

$$p_x \sim x, \quad p_y \sim y, \quad E/c^2 \sim t.$$

Replacing the indicated quantities in the Lorentz transformation (6.8), we immediately obtain the transformation of momentum and energy sought:

$$p'_x = \frac{p_x - EV/c^2}{\sqrt{1-(V/c)^2}}, \quad p'_y = p_y, \quad E' = \frac{E - p_x V}{\sqrt{1-(V/c)^2}}, \tag{7.17}$$

where V is the velocity of the K' frame relative to the K frame.

These formulae express the transformation law for the momentum and energy projections of a particle on transition from the K to K' frame.

More compact notation. At present all formulae of relativistic mechanics are customarily written in a more compact form using the following abbreviations:

(1) the quantities mc^2 and pc are denoted simply by m and p and expressed accordingly in energy units (e.g., in MeV units);

(2) all velocities are expressed in units of the velocity of light and denoted by β:

$$\beta = v/c; \tag{7.18}$$

(3) the frequently occurring factor $1/\sqrt{1-\beta^2}$ is denoted by γ, the so-called *Lorentz factor*:

$$\gamma = 1/\sqrt{1-\beta^2}. \tag{7.19}$$

These designations dramatically simplify not only the appearance of the formulae but all transformations and calculations as well. The basic formulae of relativistic dynamics in the new notation are given below:

relativistic momentum (7.3)

$$\mathbf{p} = \frac{m_0 \boldsymbol{\beta}}{\sqrt{1-\beta^2}} = \gamma m_0 \boldsymbol{\beta}, \tag{7.20}$$

kinetic (7.9) and total (7.10) energies

$$T = m_0 \left(\frac{1}{\sqrt{1-\beta^2}} - 1 \right) = m_0 (\gamma - 1), \tag{7.21}$$

$$E = m = m_0 + T = \gamma m_0; \tag{7.22}$$

relations between energy and momentum (7.12)-(7.15):

$$E^2 - p^2 = m_0^2 = \text{inv}, \tag{7.23}$$

$$\mathbf{p} = E\boldsymbol{\beta}, \tag{7.24}$$

$$p = \sqrt{T(T + 2m_0)}; \tag{7.25}$$

the Lorentz transformation of momentum and energy (7.17):

$$\left.\begin{aligned} p'_x &= \frac{p_x - \beta E}{\sqrt{1-\beta^2}} = \gamma (p_x - \beta E), \\ p'_y &= p_y, \\ E' &= \frac{E - \beta p_x}{\sqrt{1-\beta^2}} = \gamma (E - \beta p_x). \end{aligned}\right\} \tag{7.26}$$

§ 7.5. System of Relativistic Particles

About the energy and momentum of a system. Up to now we restricted ourselves to consideration of the behaviour of a single particle. In contrast to the dynamics of a single particle, the development of the dynamics of a system of particles proves to be a much more complicated task in the theory of relativity. Nevertheless, a number of important general laws can be established in this case as well.

If we wish to examine the motion of a system as a whole, then, neglecting the internal processes in the system and ignoring its spatial dimensions, that system can be regarded as a mass point (a particle). Accordingly, a system of relativistic particles can be described by the total energy E, momentum $\mathbf{p}$, and rest mass M_0, and the relations derived earlier can be considered valid for the system of particles as a whole.

We have to establish now how to interpret the total energy E, the momentum $\mathbf{p}$, and the rest mass M_0 of a system as a whole. In the general case, if the system consists of interacting relativistic particles, its total energy is

$$E = \sum m_i c^2 + W, \tag{7.27}$$

where $m_i c^2$ is the total energy of the ith particle (recall that this quantity does not include the energy of interaction with other particles), and W is the total energy of interaction of all particles of the system.

In classical mechanics W is the potential energy of interaction of a system's particles, a quantity depending only on the configuration of the system (for a given character of interaction). It turns out that in relativistic dynamics *there is no* such concept as the potential energy of interaction of particles. This is due to the fact that the very concept of potential energy is closely connected with the concept of long-range action (instantaneous interaction transmission). Being a function of the system's configuration, potential energy is defined at every moment of time by the relative disposition of the system's particles. A change in the configuration of a system must *immediately* induce a change in potential energy. Since there is no such thing in reality (interactions are transmitted with a finite velocity), the concept of potential energy of interaction cannot be introduced for a system of relativistic particles.

An expression for the interaction energy W, and therefore the total energy E, of a system of interacting relativistic particles cannot be written in the general case. The same can be said about the system's momentum since in relativistic dynamics momentum is not a quantity independent of the energy E. Things are as complicated in the case of the rest mass M_0 of the system. In the general case it is known to be the mass in a reference frame where the given mechanical system is stationary as a whole (i.e. in the C frame).

Owing to the complications mentioned above the development of the dynamics of a system of relativistic particles is restricted to a few simple cases, two of which will be examined here: a system of *non-interacting* relativistic particles and the case of two colliding particles, which is important from a practical point of view.

System of non-interacting particles. In this case the total energy E and momentum $\mathbf{p}$ possess additive properties which can be given as

$$E = \sum m_i c^2, \quad \mathbf{p} = \sum \mathbf{p}_i, \tag{7.28}$$

where m_i and $\mathbf{p}_i$ are the relativistic mass and the momentum of the ith particle of the system. Since there is no interaction in this case, the velocities of all particles are constant and consequently the total energy and the momentum of the whole system do not change with time.

Let us introduce the rest energy E_0 for a system of particles as its total energy in the C frame, where the total momentum is $\tilde{\mathbf{p}} = \sum \tilde{\mathbf{p}}_i = 0$ and the system as a whole is at rest. Thus,

$$E_0 = \sum \tilde{E}_i, \tag{7.29}$$

where $\tilde{E}_i$ is the total energy of the ith particle in the C frame. This means that the rest energy includes not only the rest energies of all particles but also their kinetic energies $\tilde{T}_i$ in the C frame: $\tilde{E}_i = m_{0i}c^2 + \tilde{T}_i$.

Obviously, the same is true for the rest mass of the system:

$$M_0 = E_0/c^2. \tag{7.30}$$

In particular, it follows from this that the rest mass of the system is not equal to the sum of the rest masses of its constituent particles:

$$M_0 > \sum m_{0i}.$$

The introduction of the rest energy and the rest mass of a system, E_0 and M_0, makes it possible to regard a system of non-interacting relativistic particles as one particle with the total energy $E = \sum m_i c^2$, the momentum $\mathbf{p} = \sum \mathbf{p}_i$, the rest mass $M_0 = E_0/c^2$, and to claim Eqs. (7.12) and (7.14) to be valid for the system of particles as well:

$$E^2 - p^2c^2 = M_0^2c^4 = \text{inv}, \tag{7.31}$$

$$\mathbf{p} = E\mathbf{V}/c^2, \tag{7.32}$$

where $\mathbf{V}$ is the velocity of the system as a whole, i.e. the velocity of the C frame. In accordance with Eq. (7.32) this velocity may be represented in the following form:

$$\mathbf{V} = (\sum \mathbf{p}_i)/(\sum m_i), \tag{7.33}$$

where m_i is the relativistic mass of the ith particle of the system. Note that Eq. (7.33) coincides in form with the corresponding non-relativistic expression (4.9) for the velocity of the system's centre of inertia.

Collision of two particles. We shall consider the collision process as proceeding in two stages: first, the formation of a compound particle A^* and then its decay into two particles that, in the general case, may differ from the initial ones:

$$A_1 + A_2 \rightarrow A^* \rightarrow A_3 + A_4 + \ldots .$$

In the process of the convergence of particles A_1 and A_2 the interaction between them may not remain weak, and Eq. (7.28) becomes inapplicable. However, after the resulting particles have separated far from each other, Eq. (7.28) becomes applicable again.

In the given case the sum of the total energies of the two initial particles (when they are so far from each other that their interaction is negligible) can be shown to equal the total energy of the compound particle. The same is true for the second stage of the process, that is, the decay. In other words, it may be shown that the total energy conservation law proves to hold true for this process in the following form:

$$E_1 + E_2 = E^* = E_3 + E_4 + \ldots . \qquad (7.34)$$

We shall demonstrate that this is really so by the following simple example.

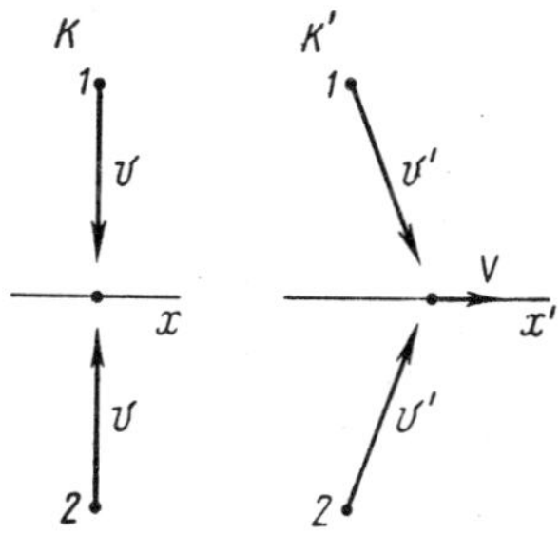

Fig. 136

Let us imagine a collision of two identical particles *1* and *2* that results in the formation of a certain compound particle. Suppose the particles move toward each other in the K frame before the collision with the same velocity v as shown in Fig. 136. Let us consider this process in the K' frame moving to the left with the velocity $\mathbf{V}$ relative to the K frame. Since in the K frame the velocity of each particle is perpendicular to the vector $\mathbf{V}$, the two particles in the K' frame, in accordance with Eq. (6.14), have x component equal to V. The compound particle formed, whose relativistic mass is denoted by M, has the same velocity in the K' frame. Applying the momentum conservation law before and after the collision (to the x component of the momentum), we obtain $2m(v')V = M \cdot V$, where v' is the velocity of each initial particle in the K' frame. Hence,

$$2m(v') = M,$$

i.e. the sum of the relativistic masses of the initial particles is equal to the relativistic mass of the formed particle. The situation is similar in the K frame. Indeed, if the value of V is small, the velocity v' is practically equal to v, and the mass M to the rest mass M_0 of the formed particle, so that in the K frame

$$2m(v) = M_0.$$

It is seen that the rest mass of the formed particle is greater than the sum of the rest masses of the initial particles. The kinetic energy of the initial particles experiences a transformation which causes the rest mass of the formed particle to exceed the sum of the rest masses of the initial particles.

Thus, we have shown that due to the system's momentum conservation the sum of the relativistic masses of the initial particles equals the relativistic mass of the formed particle. The same is obviously true for the total energy. Therefore, we can assert that the conservation of the total energy in the form described by Eq. (7.34) indeed occurs in the considered stages of that process.

As we already mentioned at the end of § 7.3, the energy conservation law, when applied to nuclear processes, made it possible to experimentally verify the validity of one of the fundamental laws of the theory of relativity, the mass-energy relationship. Let us consider some examples.

Example 1. Energy yield of nuclear reactions. Let us consider a nuclear reaction of the type

$$A_1 + A_2 \rightarrow A_3 + A_4,$$

where the initial nuclei are on the left-hand side and the reaction product nuclei on the right-hand side. We apply the law of conservation of total energy to this reaction:

$$E_1 + E_2 = E_3 + E_4.$$

Recalling that the total energy of each particle may be given as $E = m_0c^2 + T$, where m_0 is the rest mass of a nucleus and T is its kinetic energy, we rewrite the preceding equality as

$$(m_1+m_2)c^2+T_{12}=(m_3+m_4)c^2+T_{34},$$

where T_{12} and T_{34} are the total kinetic energies of nuclei before and after the reaction. Hence,

$$T_{34}-T_{12}=(m_1+m_2)c^2-(m_3+m_4)c^2.$$

The left-hand side of this equality is the increment of the overall kinetic energy of the nuclei of the given system. It is referred to as the *energy yield of a nuclear reaction* and is denoted by the letter Q. Thus,

$$Q=[(m_1+m_2)-(m_3+m_4)]c^2.$$

This quantity may have either sign depending on the nature of the nuclear reaction. Thus, the energy yield of a nuclear reaction is determined by the difference of the cumulative rest masses of nuclei before and after the reaction. All the quantities involved in this relation can be experimentally measured with a sufficiently high accuracy, verifying thereby the equality itself.

Let us consider the specific nuclear reaction

$$^7\text{Li} + {}^1\text{H} \rightarrow 2\,{}^4\text{He}.$$

The rest masses of these nuclei measured in atomic mass units (amu) are equal to 7.0160, 1.0078, and 4.0024 amu respectively. From this it is not difficult to calculate that the sum of rest masses of the nuclei decreases by 0.019 amu as a result of the reaction. Since one amu corresponds to an energy of 931.4 MeV, we find $Q = 0.019 \cdot 931.4$ MeV $= 17.7$ MeV. This value agrees very accurately with experimental data.

Example 2. Decay of a particle. Suppose a stationary particle A_1 spontaneously decays producing two particles A_2 and A_3: $A_1 \rightarrow A_2 + A_3$. In accordance with the law of conservation of total energy,

$$E_1 = E_2 + E_3.$$

As the total energy of each particle is $E = m_0c^2 + T$, the preceding equality takes the form

$$m_1c^2=(m_2+m_3)c^2+T_{23},$$

where T_{23} is the overall kinetic energy of the resulting particles. This energy is referred to as the *decay energy* Q. Thus,

$$Q=[m_1-(m_2+m_3)]c^2.$$

Since Q is essentially a positive quantity, the spontaneous decay of a particle is possible only if

$$m_1 > m_2 + m_3,$$

that is, if the rest mass of the initial particle exceeds the sum of the rest masses of the formed particles. Otherwise, spontaneous decay is impossible. Experimental evidence fully confirms this conclusion.

Let us consider, for example, the decay of a pi-meson. It is an experimental fact that charged pi-mesons disintegrate into a mu-meson and a neutrino ν: $\pi \to \mu + \nu$. The tabulated data give the rest masses of these particles (in electron rest mass units) as 273.2, 206.8 and 0 respectively. It follows that the rest mass decreases by 66.4 emu as a result of the decay. Since one emu corresponds to an energy of 0.51 MeV, the energy of this decay $Q = 66.4 \cdot 0.51$ MeV $= 34$ MeV, which accurately agrees with experimental data.

Since the collision of particles and the subsequent decay of the compound particle do not involve any change in the total energy of the system (and consequently, its momentum), another important conclusion can be inferred: for a system, the quantity $E^2 - p^2c^2$ is invariant not only with respect to different inertial reference frames but also with respect to the above-mentioned stages of a collision process.

Imagine, for example, two relativistic particles to experience a collision which leads to the generation of a new particle with rest mass M_0. If in the K frame of reference the total energies of the particles are equal to E_1 and E_2 before the collision (and their momenta to $\mathbf{p}_1$ and $\mathbf{p}_2$ respectively), we may write immediately that on transition from the K frame (*prior* to the collision) to the C frame (*after* the collision) the following equality holds:

$$\underbrace{(E_1 + E_2)^2 - (\mathbf{p}_1 + \mathbf{p}_2)^2 c^2}_{K \text{ frame}} = \underbrace{M_0^2 c^4}_{C \text{ frame}}, \tag{7.35}$$

in which it is taken into account that the formed particle is at rest in the C frame.

The invariance of the quantity $E^2 - p^2c^2$ provides us with a means to investigate the various processes of decay and collision of relativistic particles. Its application simplifies drastically both the analysis of the processes themselves and the appropriate calculations.

Example. In the K reference frame a particle possessing a rest mass m_0 and a kinetic energy T strikes a stationary particle with the same rest mass. Let us find the rest mass M_0 and the velocity V of the compound particle formed as a result of the collision.

Making use of the invariance of the quantity $E^2 - p^2c^2$, we write

$$E^2 - p^2c^2 = M_0^2c^4,$$

where the left-hand side of the equality relates to the K frame (prior to the collision) and the right-hand side to the C frame (after the collision). In this case, $E = T + 2m_0c^2$; besides, in accordance with Eq. (7.15), $p^2c^2 = T(T + 2m_0c^2)$, and therefore,

$$(T+2m_0c^2)^2 - T(T+2m_0c^2) = M_0^2c^4.$$

Whence,

$$M_0 = \sqrt{2m_0(T+2m_0c^2)}/c.$$

The velocity of the formed particle is the velocity of the C frame. In accordance with Eq. (7.32),

$$V = pc^2/E = c\sqrt{T(T+2m_0c^2)}/(T+2m_0c^2) = c/\sqrt{1+2m_0c^2/T}.$$

Problems to Chapter 7

Attention! In problems 7.4 through 7.11 we employ the abbreviated notation described at the end of § 7.4. (e.g., p and m_0 are the abbreviated forms of the quantities pc and m_0c^2).

● **7.1. Motion due to a longitudinal force.** A particle of rest mass m_0 begins moving under the action of a constant force **F**. Find the time dependence of the particle's velocity.

Solution. Multiply both sides of Eq. (7.5) by dt. Then

$$d\left(m_0v/\sqrt{1-(v/c)^2}\right) = F\,dt.$$

Integrating this expression and taking into account that $v = 0$ at the initial moment, we obtain $m_0v/\sqrt{1-(v/c)^2} = Ft$. Whence

$$v(t) = (Ft/m_0)/\sqrt{1+(Ft/m_0c)^2}.$$

Let us compare the expression thus obtained with the classical one. According to Newton's second law, $w = F/m_0$ and the velocity $v_{cl} = Ft/m_0$, and that is why the preceding expression for the velocity $v(t)$ may be presented as

$$v(t) = v_{cl}/\sqrt{1+(v_{cl}/c)^2}.$$

From this it is seen that $v < v_{cl}$, i.e. the actual velocity v of the particle grows more slowly with time as compared to v_{cl}, and the velocity $v \to c$ as $t \to \infty$ (Fig. 137).

It is interesting to note that the momentum of the particle grows linearly with time: from the equation $d\mathbf{p}/dt = \mathbf{F}$ it follows that $p = Ft$. This is a characteristic property of relativistic motion: while the velocity of a particle approaches a certain limit (i.e. becomes practically constant), the momentum of that particle keeps growing.

● **7.2. Motion due to a transverse force.** A relativistic particle of rest mass m_0 and charge q moves in a stationary uniform magnetic field whose induction is equal to **B**. The particle circumscribes a circle of radius ρ in a plane perpendicular to the vector **B**. Find the momentum and the angular rotation frequency of the particle.

Solution. In this case the particle moves due to the *Lorentz force* $\mathbf{F} = q\,[\mathbf{vB}]$, where **v** is the velocity of the particle. Since $\mathbf{F} \perp \mathbf{v}$, the magnitude of the velocity of the particle $v =$ const and Eq. (7.5) takes the form

$$m\mathbf{w} = q\,[\mathbf{vB}],$$

where m is the relativistic mass of the particle. Recalling that **w** is a normal acceleration whose magnitude is equal to v^2/ρ, we rewrite the preceding equation as $mv^2/\rho = qvB$. Hence, the momentum of the particle is

$$p = mv = q\rho B. \tag{1}$$

Thus, the product ρB may serve as a measure of the relativistic momentum of the given particle.

With allowance made for Eq. (1) the angular frequency of the particle is

$$\omega = v/\rho = p/m\rho = qB/m.$$

It follows that the angular frequency ω depends on the velocity of the particle: the greater the velocity of the particle, and therefore,

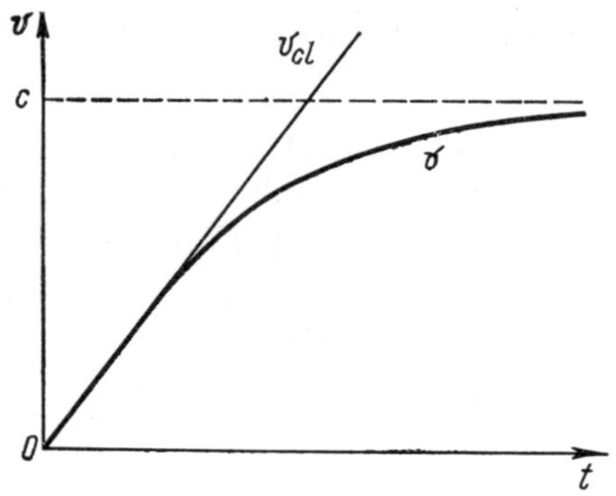

Fig. 137

the relativistic mass m, the lower the angular frequency ω. However, at low velocities ($v \ll c$) $m \to m_0$, and

$$\omega = qB/m_0 = \text{const},$$

i.e. in this velocity range the frequency ω is practically independent of the velocity.

● **7.3.** At the moment $t = 0$ a relativistic proton with momentum $\mathbf{p}_0$ flies into a region where there is a transverse uniform electric field of strength $\mathbf{E}$, with $\mathbf{p}_0 \perp \mathbf{E}$. Find the time dependence of the angle θ at which the proton is deflected from the initial direction of its motion.

Solution. Taking the x coordinate along the vector $\mathbf{p}_0$ and the y coordinate along $\mathbf{E}$, we write Eq. (7.4) in projections on these axes:

$$dp_x/dt = 0, \qquad dp_y/dt = eE,$$

where e is the proton charge. From these equations it follows that

$$p_x = p_0, \qquad p_y = eEt,$$

or

$$m_0 v_x/\sqrt{1-(v/c)^2} = p_0, \qquad m_0 v_y/\sqrt{1-(v/c)^2} = eEt. \tag{1}$$

From the ratio of the last two equalities we get

$$\tan\theta = v_y/v_x = eEt/p_0.$$

It is interesting to point out that in contrast to the non-relativistic case v_x decreases with an increase in time here. To make sure of this, let us square both equalities (1) and then add separately their left-hand and right-hand sides:

$$\frac{m_0^2 (v_x^2 + v_y^2)}{1-(v/c)^2} = p_0^2 + (eEt)^2.$$

Recalling that $v_x^2 + v_y^2 = v^2$, we obtain

$$\left(\frac{v}{c}\right)^2 = \left[1 + \frac{m_0^2 c^2}{p_0^2 + (eEt)^2}\right]^{-1}.$$

Substituting this expression into the first equality of (1), we get

$$v_x = c/\sqrt{1 + (m_0 c/p_0)^2 + (eEt/p_0)^2},$$

i.e. v_x really decreases in the course of time t.

● **7.4. Symmetric elastic scattering.** A relativistic proton possessing kinetic energy T collides elastically with a stationary proton with the result that both protons move apart symmetrically relative to the initial motion direction. Find the angle between the motion directions of the protons after the collision.

Solution. In symmetric scattering of the protons their momenta and energies must be equal in magnitude. This is immediately seen from the triangle of momenta (Fig. 138), which expresses the momentum conservation law. From that triangle we can write, in accordance with the cosine theorem,

$$p^2 = 2p'^2 + 2p'^2 \cos\theta,$$

whence

$$\cos\theta = p^2/2p'^2 - 1.$$

Using Eq. (7.25) and taking into account that $T = 2T'$, where T'

is the kinetic energy of each proton after the collision, we find

$$\frac{p^2}{p'^2}=\frac{T(T+2m_0)}{T'(T'+2m_0)}=4\frac{T+2m_0}{T+4m_0},$$

where m_0 is the rest mass of a proton. Substituting this expression into the formula for $\cos\Theta$, we obtain

$$\cos\Theta = T/(T+4m_0).$$

Note that as distinct from the non-relativistic case when $\Theta = \pi/2$, here $\Theta < \pi/2$.

● **7.5.** A photon of energy ε is scattered by a stationary free electron. Find the energy ε' of the scattered photon if the angle between the motion directions of the incoming photon and the scattered one is equal to θ.

Solution. Let us apply the momentum and energy conservation laws to the given process:

$$T_e = \varepsilon - \varepsilon', \quad \mathbf{p}_e = \mathbf{p} - \mathbf{p}',$$

where T_e and $\mathbf{p}_e$ are the kinetic energy and the momentum of the recoiled electron, and $\mathbf{p}$ and $\mathbf{p}'$ are the momenta of the incoming

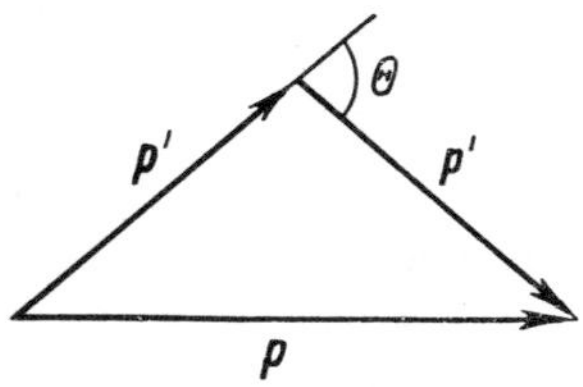

Fig. 138

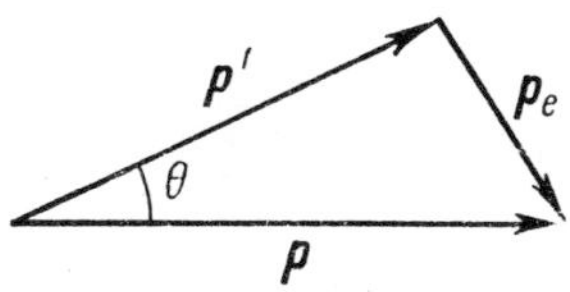

Fig. 139

and scattered photons. According to the cosine theorem it follows from the triangle of momenta (Fig. 139) that

$$p_e^2 = p^2 + p'^2 - 2pp'\cos\theta.$$

Substituting here $p = \varepsilon$, $p' = \varepsilon'$ and $p_e = \sqrt{T_e(T_e + 2m_e)} = \sqrt{(\varepsilon - \varepsilon')(\varepsilon - \varepsilon' + 2m_e)}$, where m_e is the rest mass of the electron, we obtain after simple transformations

$$\varepsilon' = \frac{\varepsilon}{1+(2\varepsilon/m_e)\sin^2(\theta/2)}.$$

◑ **7.6.** Two protons move toward each other with equal kinetic energies T (in the K reference frame). Find the kinetic energy T' of one proton with respect to the other.

Solution. Let us take advantage of the invariance of the quantity $E^2 - p^2$, writing it in the K frame (which is also the C frame here) and in the reference frame fixed to one of the protons

$$[2(T+m_p)]^2=(T'+2m_p)^2-T'(T'+2m_p),$$

where m_p is the rest mass of a proton. From this it follows that

$$T'=2T(T+2m_p)/m_p.$$

For example, in the case of protons ($m_p \simeq 1$ GeV), if $T = 50$ GeV), then $T' = 5\cdot 10^3$ GeV. The possibility of such a large energy "gain" underlies the method of head-on collision beams.

7.7. Energy diagram of a nuclear reaction. A particle A_1 with kinetic energy T_1 strikes a stationary nucleus A_2 (in the K frame). As a result of the reaction the nuclei A_3 and A_4 are formed:

$$A_1 + A_2 \to A_3 + A_4.$$

The rest masses of the particles are equal to m_1, m_2, m_3, and m_4 respectively. Illustrate the energy level diagram of the nuclear reaction for two cases: (a) $m_1 + m_2 > m_3 + m_4$, and (b) $m_1 + m_2 < m_3 + m_4$. For the second case find the threshold kinetic energy T_{1thr} of the incoming particle in the K frame.

Solution. From the law of conservation of the total energy it follows that in the C frame

$$\tilde{T}_{12}+m_1+m_2=\tilde{T}_{34}+m_3+m_4,$$

where $\tilde{T}_{12}$ and $\tilde{T}_{34}$ are the overall kinetic energies of the particles

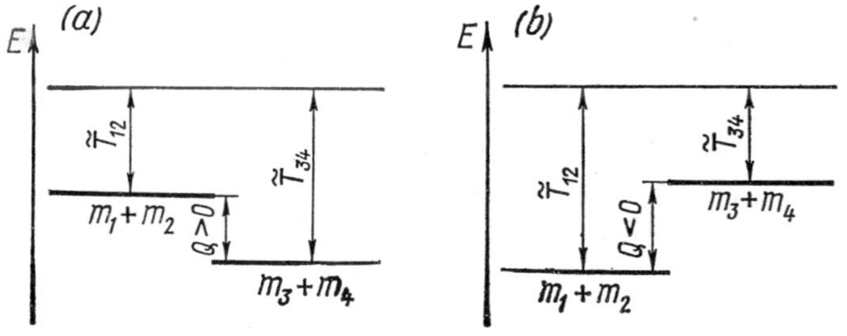

Fig. 140

before and after the reaction. Denoting the increment of the kinetic energy $\tilde{T}_{34}-\tilde{T}_{12}$ by Q, we may write the preceding expression as

$$Q = (m_1 + m_2) - (m_3 + m_4),$$

where Q is the energy yield of the nuclear reaction. The energy diagram of the reaction is illustrated in Fig. 140 for both cases. In case (a) the effect is positive, $Q > 0$: the overall kinetic energy increases at the expense of a decrease in the sum of the rest masses of the system's particles; in case (b) the opposite is true.

In the latter case, as it is seen from Fig. 140*b*, the nuclear reaction is possible only if $\tilde{T}_{12} \geqslant |Q|$. Here the equality sign corresponds to the threshold value of the energy $\tilde{T}_{12}$. In accordance with Eq. (4.16), at low velocities,

$$T_{12thr} = \frac{\mu v_{rel}^2}{2} = \frac{m_2}{m_1+m_2}\frac{m_1 v_1^2}{2} = \frac{m_2}{m_1+m_2} T_{1thr} = |Q|,$$

whence

$$T_{1thr} = |Q|(m_1+m_2)/m_2.$$

● **7.8. Threshold energy** (the minimum energy required to activate a given process).

1. A relativistic particle of rest mass m_0 strikes a stationary particle of rest mass M_0. As a result of the impact, particles of rest masses m_1, m_2, ... are generated according to the scheme

$$m_0 + M_0 \rightarrow m_1 + m_2 + \ldots .$$

Find the threshold kinetic energy T_{thr} of the incoming particle.

2. Find the threshold energy of a photon for electron-positron pair production in the field of a stationary proton.

Solution. 1. First of all, it is clear that threshold energy is meaningful only when the sum of the rest masses of the initial particles is less than that of the particles produced. To find T_{thr}, let us make use of the invariance of the quantity $E^2 - p^2$. Let us write this quantity prior to the collision for $T = T_{thr}$ in the reference frame where the particle M_0 is at rest, and after the collision in the C frame:

$$E^2 - p^2 = \tilde{E}^2,$$

or

$$(T_{thr} + m_0 + M_0)^2 - T_{thr}(T_{thr} + 2m_0) = (m_1 + m_2 + \ldots)^2.$$

Here we have taken into account that in the C frame the kinetic energy of the formed particles is equal to zero at the threshold of the reaction, and therefore their total energy equals the sum of the rest masses of the individual particles. From the latter equation we get

$$T_{thr} = [(m_1 + m_2 + \ldots)^2 - (m_0 + M_0)^2]/2M_0.$$

2. Let us write $E^2 - p^2$ before the interaction in the reference frame where the proton is at rest and after the interaction in the C frame. At the threshold value of the energy ε of the incoming photon

$$(\varepsilon_{thr} + M_0)^2 - \varepsilon_{thr}^2 = (M_0 + 2m_0)^2,$$

where M_0 is the rest mass of a proton, and m_0 is the rest mass of an electron (positron). Hence,

$$\varepsilon_{thr} = 2m_0(1 + m_0/M_0).$$

It is seen that pair production requires the photon energy to exceed $2m_0$.

● **7.9. Decay of a moving particle.** A relativistic π^0 meson of rest mass m_0 disintegrates during its motion into two gamma photons with energies ε_1 and ε_2 (in the K reference frame). Find the angle Θ of divergence of the gamma photons.

Solution. Using the invariance of the expression $E^2 - p^2$, we write it in the C frame before the decay and in the K frame after the decay:

$$m_0^2=(\varepsilon_1+\varepsilon_2)^2-(\mathbf{p}_1+\mathbf{p}_2)^2,$$

where $\mathbf{p}_1$ and $\mathbf{p}_2$ are the momenta of the gamma photons. We transform the right-hand side of this equation taking into account that $p_1 = \varepsilon_1$ and $p_2 = \varepsilon_2$; then

$$m_0^2=2\varepsilon_1\varepsilon_2-2\mathbf{p}_1\mathbf{p}_2, \quad \text{or} \quad m_0^2=2\varepsilon_1\varepsilon_2(1-\cos\Theta).$$

Hence

$$\sin(\Theta/2)=m_0/2\sqrt{\varepsilon_1\varepsilon_2}.$$

● **7.10.** The total momentum and energy of a system of two non-interacting particles are $\mathbf{p} = \mathbf{p}_1 + \mathbf{p}_2$ and $E = E_1 + E_2$. Demonstrate explicitly that the Lorentz transformation for the total momentum $\mathbf{p}$ and energy E is consistent with the invariance of the quantity $E^2 - p^2$ for the given system.

Solution. Using the Lorentz transformation for momentum and energy (7.26), we find the projections of the total momentum and energy in another (primed) reference frame possessing the velocity β and the corresponding Lorentz factor γ:

$$p'_x=p'_{1x}+p'_{2x}=\gamma(p_{1x}+p_{2x})-\gamma\beta(E_1+E_2)=\gamma(p_x-\beta E);$$

$$p'_y=p'_{1y}+p'_{2y}=p_{1y}+p_{2y}=p_y;$$

$$E'=E'_1+E'_2=\gamma(E_1+E_2)-\gamma\beta(p_{1x}+p_{2x})=\gamma(E-\beta p_x).$$

Hence

$$E'^2-p'^2=E'^2-(p_x'^2+p_y'^2)=E^2-p^2.$$

● **7.11.** In the laboratory reference frame a photon with energy ε strikes a stationary particle A of rest mass m_0. Find:

(1) the velocity of the C frame for these two particles;

(2) the energies of the photon and the particle in the C frame.

Solution. 1. In accordance with Eq. (7.32) the velocity of the C frame is

$$\beta = p/E = \varepsilon/(\varepsilon + m_0).$$

2. From the Lorentz transformation for energy (7.26) it follows that the photon energy in the C frame is

$$\tilde{\varepsilon}=\gamma(\varepsilon-\beta p)=\gamma(\varepsilon-\beta\varepsilon)=\varepsilon\frac{1-\beta}{\sqrt{1-\beta^2}}=\varepsilon\sqrt{\frac{1-\beta}{1+\beta}}.$$

Substituting the expression for β from the previous part, we obtain

$$\tilde{\varepsilon}=\varepsilon\sqrt{m_0/(2\varepsilon+m_0)}.$$

The energy of the particle A in the C frame is

$$\tilde{E}_A = m_0/\sqrt{1-\beta^2} = (\varepsilon + m)\sqrt{m_0/(2\varepsilon + m_0)}.$$

The correctness of the formulae obtained may be checked by making use of the invariance of the expression $E^2 - p^2$ on transition from the laboratory reference frame to the C frame:

$$(\varepsilon + m_0)^2 - \varepsilon^2 = (\tilde{\varepsilon} + \tilde{E}_A)^2.$$

1. Motion of a Point in Polar Coordinates

In polar coordinates ρ, φ the position of a point A on a plane is defined if we know its distance ρ from the origin O (Fig. 141*a*) and the angle φ between the radius vector $\boldsymbol{\rho}$ of the point and a chosen direction OO', the zero reading of the angular coordinate φ.

Let us introduce unit vectors $\mathbf{e}_\rho$ and $\mathbf{e}_\varphi$ associated with the moving point A and oriented in the direction of the increasing coordinates ρ and φ as shown in Fig. 141*a*. Unlike the unit vectors

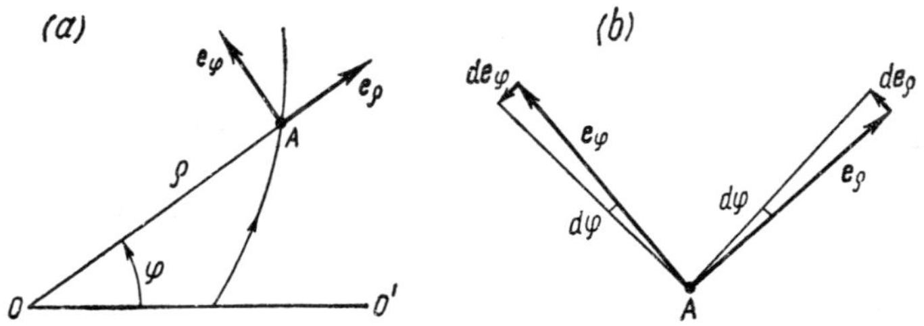

Fig. 141

of the Cartesian system of coordinates, $\mathbf{e}_\rho$ and $\mathbf{e}_\varphi$ are *movable*, that is, they change their direction as the point A moves. Let us find their time derivatives, which will be required later. During the motion of the point A both unit vectors turn in the same direction through the same angle $d\varphi$ in the time interval dt (Fig. 141*b*) and acquire the increments:

$$d\mathbf{e}_\rho = 1\cdot d\varphi\cdot\mathbf{e}_\varphi;\quad d\mathbf{e}_\varphi = 1\cdot d\varphi\cdot(-\mathbf{e}_\rho).$$

Dividing both expressions by dt, we obtain

$$\dot{\mathbf{e}}_\rho = \dot{\varphi}\mathbf{e}_\varphi;\quad \dot{\mathbf{e}}_\varphi = -\dot{\varphi}\mathbf{e}_\rho, \tag{1}$$

where a dot over a letter signifies differentiation with respect to time.

Let us now determine the velocity and acceleration of the point A, writing its radius vector $\boldsymbol{\rho}$ in the form

$$\boldsymbol{\rho} = \rho\mathbf{e}_\rho. \tag{2}$$

The velocity **v** *of a point.* Let us differentiate Eq. (2) with respect to time, allowing for Eq. (1):

$$\mathbf{v} = \dot{\rho}\mathbf{e}_\rho + \rho\dot{\varphi}\mathbf{e}_\varphi, \tag{3}$$

i.e. the projections of the vector **v** on the movable unit vectors $\mathbf{e}_\rho$ and $\mathbf{e}_\varphi$ are equal to

$$v_\rho = \dot{\rho}; \quad v_\varphi = \rho\dot{\varphi}, \tag{4}$$

and the magnitude of the velocity vector is $v = \sqrt{\dot{\rho}^2 + \rho^2\dot{\varphi}^2}$.

The acceleration **w** *of a point.* Differentiating Eq. (3) with respect to time once again, we get

$$\mathbf{w} = \frac{d\mathbf{v}}{dt} = \ddot{\rho}\mathbf{e}_\rho + \dot{\rho}\dot{\mathbf{e}}_\rho + \frac{d}{dt}(\rho\dot{\varphi})\,\mathbf{e}_\varphi + \rho\dot{\varphi}\dot{\mathbf{e}}_\varphi.$$

After simple transformations we find, taking into account Eq. (1):

$$\mathbf{w} = (\ddot{\rho} - \rho\dot{\varphi}^2)\,\mathbf{e}_\rho + (2\dot{\rho}\dot{\varphi} + \rho\ddot{\varphi})\,\mathbf{e}_\varphi, \tag{5}$$

i.e. the projections of the vector **w** on the unit vectors $\mathbf{e}_\rho$ and $\mathbf{e}_\varphi$ are equal to

$$w_\rho = \ddot{\rho} - \rho\dot{\varphi}^2, \tag{6}$$

$$w_\varphi = 2\dot{\rho}\dot{\varphi} + \rho\ddot{\varphi} = \frac{1}{\rho}\frac{d}{dt}(\rho^2\dot{\varphi}).$$

The fundamental equation of dynamics in polar coordinates. The fundamental equation of dynamics $m\mathbf{w} = \mathbf{F}$ in projections on the movable unit vectors $\mathbf{e}_\rho$ and $\mathbf{e}_\varphi$ is easy to obtain at once, making use of Eqs. (6):

$$\left.\begin{aligned} m(\ddot{\rho} - \rho\dot{\varphi}^2) &= F_\rho, \\ m\frac{1}{\rho}\frac{d}{dt}(\rho^2\dot{\varphi}) &= F_\varphi \end{aligned}\right\} \tag{7}$$

where F_ρ and F_φ are the projections of the vector **F** on the unit vectors $\mathbf{e}_\rho$ and $\mathbf{e}_\varphi$ (Fig. 142). In that figure $F_\rho < 0$ and $F_\varphi > 0$.

2. On Keplerian Motion

The motion of a particle in a central field of forces that are inversely proportional to the square of the distance from the field centre is called Keplerian. The Newtonian attraction forces between mass points (or bodies possessing spherical symmetry) and Coulomb forces between point charges are forces of this kind.

In such a field the potential energy of a particle is $U = -\alpha/\rho$, where α is a constant and ρ is the distance from the field centre. Let us examine the case when $\alpha > 0$, i.e. the force acting on a particle of mass m is directed toward the field centre (attraction). What shape does the trajectory of the particle have in polar coordinates $\rho(\varphi)$ if $\rho(0) = \rho_0$ at $\varphi = 0$ and its velocity is perpendicular to the radius vector and is equal to v_0 (Fig. 143)?

To solve this problem, the laws of conservation of energy and angular momentum are usually utilized. In polar coordinates these laws yield

$$\frac{m}{2}(\dot{\rho}^2+\rho^2\dot{\varphi}^2)-\frac{\alpha}{\rho}=E; \qquad m\rho^2\dot{\varphi}=L,$$

where E and L are the total mechanical energy and the angular momentum of the particle relative to the point O, the field centre. Both of these quantities are easy to find from the initial conditions.

These equations are solved as follows. Initially, in the first equation differentiation with respect to time is replaced by differentiation

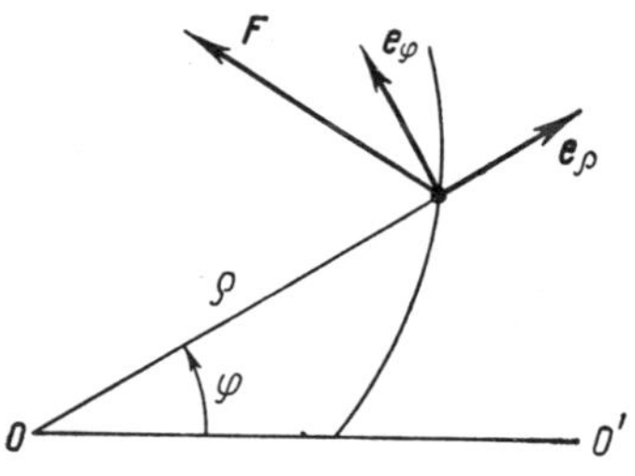

Fig. 142

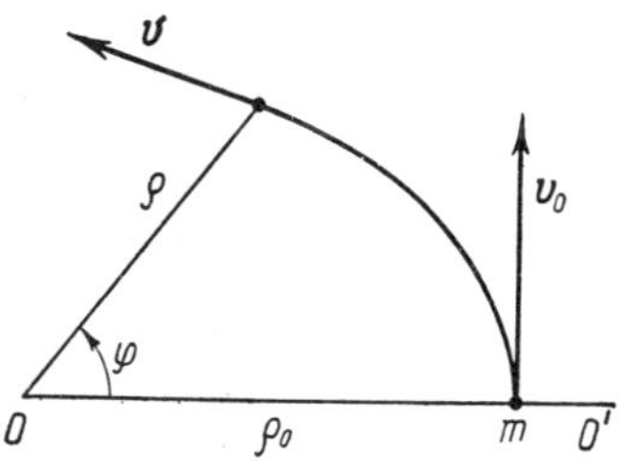

Fig. 143

with respect to φ; this can be done by using the second equation: $dt = (m\rho^2/L)\,d\varphi$. Then the variables ρ, φ are separated, i.e. the obtained expression is reduced to the form $d\varphi = f(\rho)\,d\rho$. And finally, that equation is integrated, with account taken of the initial conditions. The result of integration yields the sought solution $\rho(\varphi)$.

We shall not describe the rather cumbersome procedure for solving these equations here. If necessary, it may be found in almost any textbook on theoretical physics or mechanics. We shall restrict ourselves to an analysis of the solution obtained, which has the form

$$\rho(\varphi) = \rho_0/[a + (1-a)\cos\varphi], \tag{1}$$

where $a = \alpha/m\rho_0 v_0^2$.

It is known from mathematics that Eq. (1) describes a curve of the second order. Depending on the value of the parameter a this may be an ellipse (circle), a parabola, or a hyperbola.

1. It is immediately seen that at $a = 1$ ρ does not depend on φ, i.e. the trajectory is a circle. A particle has such a trajectory at a velocity v_0 equal to

$$v_{\mathrm{I}} = \sqrt{\alpha/m\rho_0}. \tag{2}$$

2. For all values of the parameter a at which ρ is finite up to $\varphi = \pi$, the trajectory has the form of an ellipse. At $\varphi = \pi$, as it follows from Eq. (1),

$$\rho(\pi) = \rho_0/(2a - 1).$$

It is seen from this that $\rho(\pi)$ is finite only when $2a > 1$, i.e. when the velocity $v_0 < v_{II}$, where

$$v_{II} = \sqrt{2\alpha/m\rho_0}. \tag{3}$$

3. If $2a = 1$, i.e. $v_0 = v_{II}$, the ellipse degenerates into a parabola, which means that the particle does not come back again.

4. At $v_0 > v_{II}$ the trajectory has the shape of a hyperbola.

All of these cases are illustrated in Fig. 144. It should be pointed out that in elliptical orbits the field centre coincides with one of the

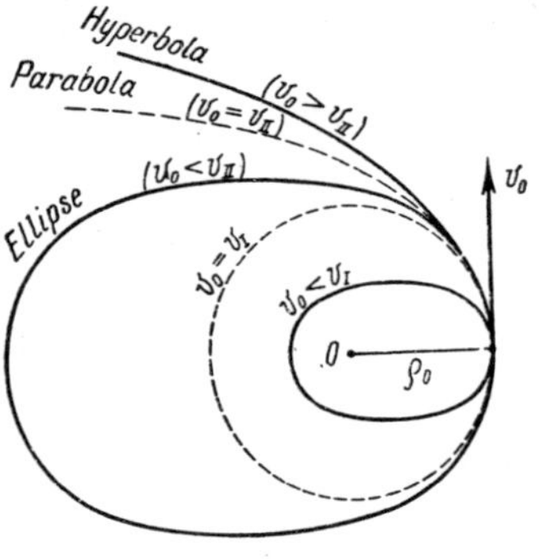

Fig. 144

ellipse's focal points: namely, with the back focus if $v_0 < v_I$, and with the front focus if $v_0 > v_I$.

Note that Eq. (1) describes, for example, the trajectories of the planets of the solar system, with $\alpha = \gamma mM$, where M is the mass of the Sun. As applied to the motion of space vehicles, v_I and v_{II} are the orbital and escape velocities respectively. Obviously, their magnitudes depend on the mass of the body that is the source of the field.

3. Demonstration of Steiner's Theorem

Theorem: the moment of inertia I of a solid body relative to an arbitrary axis z equals the moment of inertia I_C of that body relative to the axis z_C parallel to the given one and passing through the body's centre of inertia, plus the product of the mass m of the body and the square of the distance a between the axes:

$$I = I_C + ma^2.$$

Proof. Let us draw through the ith element of the body a plane perpendicular to the z axis, and in this plane, three vectors $\boldsymbol{\rho}_i$, $\boldsymbol{\rho}'_i$ and a (Fig. 145). The first two vectors describe the position of the ith element of the body relative to the z and z_C axes while the vector **a** specifies the position of the z_C axis relative to the z axis. Taking

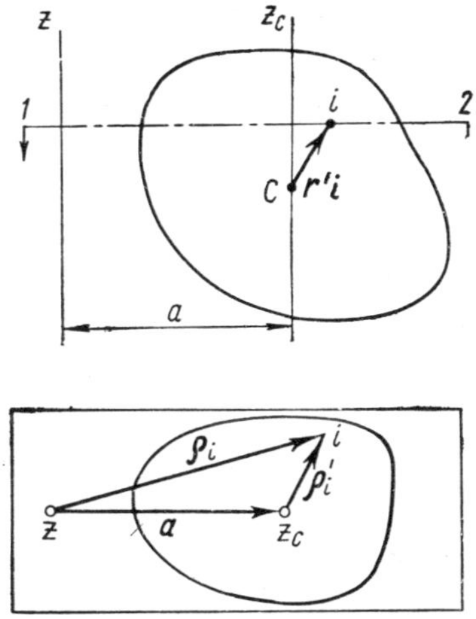

Fig. 145

advantage of the relationship between these vectors ($\boldsymbol{\rho}_i = \boldsymbol{\rho}'_i + \mathbf{a}$), we transform the expression for the moment of inertia of the body relative to the z axis:

$$I = \sum m_i\rho_i^2 = \sum m_i (\boldsymbol{\rho}'_i + \mathbf{a})^2 = \sum m_i\rho_i'^2 + 2\mathbf{a}\sum m_i\boldsymbol{\rho}'_i + \sum m_i a^2.$$

The first term on the right-hand side of that equality is the moment of inertia I_C of the body relative to the z_C axis, and the last term is equal to ma^2. What is left is to show that the middle term equals zero.

Suppose $\mathbf{r}'_i$ is the radius vector of the ith element of the body relative to the centre of inertia; then the vector $\sum m_i\mathbf{r}'_i = 0$ relative to that centre. But $\boldsymbol{\rho}'_i$ is the *vector projection* of the vector $\mathbf{r}'_i$ on the plane perpendicular to the z axis. Hence it is clear that if the composite vector is equal to zero, the sum of its vector projections on any plane is also equal to zero, i.e. $\sum m_i\boldsymbol{\rho}'_i = 0$. The theorem is thus proved.

4. Greek Alphabet

Α, α — Alpha Β, β — Beta Γ, γ — Gamma Δ, δ — Delta Ε, ε — Epsilon Ζ, ζ — Zeta Η, η — Eta Θ, θ, ϑ — Theta	Ι, ι — Iota Κ, ϰ — Kappa Λ, λ — Lambda Μ, μ — Mu Ν, ν — Nu Ξ, ξ — Xi Ο, ο — Omicron Π, π — Pi	Ρ, ρ — Rho Σ, σ — Sigma Τ, τ — Tau Υ, υ — Upsilon Φ, φ — Phi Χ, χ — Chi Ψ, ψ — Psi Ω, ω — Omega

5. Some Formulae of Algebra and Trigonometry

The roots of the quadratic equation $ax^2+bx+c=0$

$$x_{1,2}=\frac{-b\pm\sqrt{b^2-4ac}}{2a}.$$

Some approximate formulae. If $\alpha \ll 1$, then

$$(1\pm\alpha)^n=1\pm n\alpha \qquad \sin\alpha=\alpha$$

$$e^{\alpha}=1+\alpha \qquad \cos\alpha=1-\alpha^2/2$$

$$\ln(1+\alpha)=\alpha \qquad \tan\alpha=\alpha$$

The basic formulae of trigonometry

$$\sin^2\alpha+\cos^2\alpha=1$$

$$\sec^2\alpha-\tan^2\alpha=1$$

$$\csc^2\alpha-\cot^2\alpha=1$$

$$\sin\alpha\cdot\sec\alpha=1$$

$$\cos\alpha\cdot\csc\alpha=1$$

$$\tan\alpha\cdot\cot\alpha=1$$

$$\sin\alpha=\frac{1}{\sqrt{1+\cot^2\alpha}}$$

$$\cos\alpha=\frac{1}{\sqrt{1+\tan^2\alpha}}$$

$$\sin(\alpha\pm\beta)=\sin\alpha\cos\beta\pm\cos\alpha\sin\beta$$

$$\cos(\alpha\pm\beta)=\cos\alpha\cos\beta\mp\sin\alpha\sin\beta$$

$$\tan(\alpha\pm\beta)=\frac{\tan\alpha\mp\tan\beta}{1\mp\tan\alpha\cdot\tan\beta}$$

$$\cot(\alpha\pm\beta)=\frac{\cot\alpha\cdot\cot\beta\mp1}{\cot\beta\mp\cot\alpha}$$

$$\sin\alpha+\sin\beta=2\sin\frac{\alpha+\beta}{2}\cos\frac{\alpha-\beta}{2}$$

$$\sin\alpha-\sin\beta=2\cos\frac{\alpha+\beta}{2}\sin\frac{\alpha-\beta}{2}$$

Concluded

$\sin 2\alpha = 2\sin\alpha\cdot\cos\alpha$ $\cos 2\alpha = \cos^2\alpha - \sin^2\alpha$	$\cos\alpha + \cos\beta =$ $= 2\cos\frac{\alpha+\beta}{2}\cos\frac{\alpha-\beta}{2}$
$\tan 2\alpha = \frac{2\tan\alpha}{1-\tan^2\alpha}$ $\cot 2\alpha = \frac{\cot^2\alpha - 1}{2\cot\alpha}$	$\cos\alpha - \cos\beta =$ $= -2\sin\frac{\alpha+\beta}{2}\sin\frac{\alpha-\beta}{2}$ $\tan\alpha \pm \tan\beta = \frac{\sin(\alpha\pm\beta)}{\cos\alpha\cdot\cos\beta}$ $\cot\alpha \pm \cot\beta = \pm\frac{\sin(\alpha\pm\beta)}{\sin\alpha\cdot\sin\beta}$
$\sin\frac{\alpha}{2} = \sqrt{\frac{1-\cos\alpha}{2}}$ $\cos\frac{\alpha}{2} = \sqrt{\frac{1+\cos\alpha}{2}}$	$2\sin\alpha\cdot\sin\beta = \cos(\alpha-\beta) - \cos(\alpha+\beta)$ $2\cos\alpha\cdot\cos\beta = \cos(\alpha-\beta) + \cos(\alpha+\beta)$ $2\sin\alpha\cdot\cos\beta = \sin(\alpha-\beta) + \sin(\alpha+\beta)$

6. Table of Derivatives and Integrals

Function	Derivative	Function	Derivative
x^n	nx^{n-1}	$\sin x$	$\cos x$
$\frac{1}{x}$	$-\frac{1}{x^2}$	$\cos x$	$-\sin x$
$\frac{1}{x^n}$	$-\frac{n}{x^{n+1}}$	$\tan x$	$\frac{1}{\cos^2 x}$
$\sqrt{x}$	$\frac{1}{2\sqrt{x}}$	$\cot x$	$-\frac{1}{\sin^2 x}$
e^x	e^x	$\arcsin x$	$\frac{1}{\sqrt{1-x^2}}$
e^{nx}	ne^{nx}	$\arccos x$	$-\frac{1}{\sqrt{1-x^2}}$
a^x	$a^x \ln a$		

Concluded

Function	Derivative	Function	Derivative
$\ln x$	$\frac{1}{x}$	$\arctan x$	$\frac{1}{1+x^2}$
$\sqrt{u}$	$\frac{u'}{2\sqrt{u}}$	$\operatorname{arccot} x$	$-\frac{1}{1+x^2}$
$\ln u$	$\frac{u'}{u}$	$\sinh x$	$\cosh x$
		$\cosh x$	$-\sinh x$
$\frac{u}{v}$	$\frac{vu'-v'u}{v^2}$	$\tanh x$	$\frac{1}{\cosh^2 x}$
		$\coth x$	$-\frac{1}{\sinh^2 x}$

$$\int x^n\,dx=\frac{x^{n+1}}{n+1},\quad n\neq -1$$

$$\int \frac{dx}{x}=\ln x$$

$$\int \sin x\,dx=-\cos x$$

$$\int \cos x\,dx=\sin x$$

$$\int \tan x\,dx=-\ln\cos x$$

$$\int \cot x\,dx=\ln\sin x$$

$$\int \frac{dx}{\cos^2 x}=\tan x$$

$$\int \frac{dx}{\sin^2 x}=-\cot x$$

$$\int e^x\,dx=e^x$$

$$\int \frac{dx}{1+x^2}=\text{arc tan}\,x$$

$$\int \frac{dx}{\sqrt{1-x^2}}=\arcsin x$$

$$\int \frac{dx}{\sqrt{x^2-1}}=\ln\left(x+\sqrt{x^2-1}\right)$$

7. Some Facts About Vectors

Scalar product:

$$\mathbf{ab}=\mathbf{ba}=ab\cos\alpha;$$
$$\mathbf{a}(\mathbf{b}+\mathbf{c})=\mathbf{ab}+\mathbf{ac}.$$

Cross product:

$$[\mathbf{ab}]=-[\mathbf{ba}];\quad |[\mathbf{ab}]|=ab\sin\alpha;$$
$$[\mathbf{a},\ \mathbf{b}+\mathbf{c}]=[\mathbf{ab}]+[\mathbf{ac}].$$

Concluded

Mixed, or vector-scalar, product of three vectors is a scalar equal numerically to the volume of the parallelepiped constructed from these vectors:

$$\mathbf{a}\,[\mathbf{bc}]=\mathbf{b}\,[\mathbf{ca}]=\mathbf{c}\,[\mathbf{ab}];$$

$$\mathbf{a}\,[\mathbf{bc}]=-\mathbf{b}\,[\mathbf{ac}]=-\mathbf{a}\,[\mathbf{cb}].$$

Double vector product:

$$[\mathbf{a}\,[\mathbf{bc}]]=\mathbf{b}\,(\mathbf{ac})-\mathbf{c}\,(\mathbf{ab}).$$

Products of vectors in coordinate form. If

$$\mathbf{a}=a_1\mathbf{e}_1+a_2\mathbf{e}_2+a_3\mathbf{e}_3,$$

$$\mathbf{b}=b_1\mathbf{e}_1+b_2\mathbf{e}_2+b_3\mathbf{e}_3,$$

where $\mathbf{e}_1$, $\mathbf{e}_2$, $\mathbf{e}_3$ are coordinate unit vectors which are mutually perpendicular and form a right triad, then

$$\mathbf{ab}=a_1b_1+a_2b_2+a_3b_3;$$

$$[\mathbf{ab}]=\begin{vmatrix}\mathbf{e}_1 & \mathbf{e}_2 & \mathbf{e}_3\\ a_1 & a_2 & a_3\\ b_1 & b_2 & b_3\end{vmatrix}=(a_2b_3-a_3b_2)\,\mathbf{e}_1+(a_3b_1-a_1b_3)\,\mathbf{e}_2+(a_1b_2-a_2b_1)\,\mathbf{e}_3.$$

The rules for differentiating vectors depending on a certain scalar variable t:

$$\frac{d}{dt}(\mathbf{a}+\mathbf{b})=\frac{d\mathbf{a}}{dt}+\frac{d\mathbf{b}}{dt};$$

$$\frac{d}{dt}(\alpha\mathbf{a})=\frac{d\alpha}{dt}\mathbf{a}+\alpha\frac{d\mathbf{a}}{dt};$$

$$\frac{d}{dt}(\mathbf{ab})=\frac{d\mathbf{a}}{dt}\mathbf{b}+\mathbf{a}\frac{d\mathbf{b}}{dt};$$

$$\frac{d}{dt}[\mathbf{ab}]=\left[\frac{d\mathbf{a}}{dt}\mathbf{b}\right]+\left[\mathbf{a}\frac{d\mathbf{b}}{dt}\right].$$

Gradient of the scalar function φ:

$$\nabla\varphi=\frac{\partial\varphi}{\partial x}\mathbf{i}+\frac{\partial\varphi}{\partial y}\mathbf{j}+\frac{\partial\varphi}{\partial z}\mathbf{k},$$

where $\mathbf{i}$, $\mathbf{j}$, $\mathbf{k}$ are the coordinate unit vectors of the x, y, z axes.

8. Units of Mechanical Quantities in the SI and CGS Systems

Quantity	Unit		Conversion factor, 1 SI unit/1 CGS unit
	SI	CGS	
Length	m	cm	10^2
Time	s	s	1
Angle	rad	rad	1
Area	m^2	cm^2	10^4
Volume	m^3	cm^3	10^6
Velocity	m/s	cm/s	10^2
Acceleration	m/s^2	cm/s^2	10^2
Frequency	Hz	Hz	1
Angular frequency	rad/s	rad/s	1
Angular velocity	rad/s	rad/s	1
Angular acceleration	rad/s^2	rad/s^2	1
Mass	kg	g	10^3
Density	kg/m^3	g/cm^3	10^{-3}
Force	N	dyn	10^5
Pressure	Pa	dyn/cm^2	10
Work, energy	J	erg	10^7
Power	W	erg/s	10^7
Momentum	kg·m/s	g·cm/s	10^5
Power impulse	N·s	dyn·s	10^5
Force moment (torque)	N·m	dyn·cm	10^7
Angular momentum	$kg \cdot m^2/s$	$g \cdot cm^2/s$	10^7
Moment of inertia	$kg \cdot m^2$	$g \cdot cm^2$	10^7
Torque momentum	N·m·s	dyn·cm·s	10^7
Energy flux	W	erg/s	10^7
Energy flux density	W/m^2	$erg/(s \cdot cm^2)$	10^3

9. Decimal Prefixes for the Names of Units

T tera (10^{12})	c centi (10^{-2})
G giga (10^9)	m milli (10^{-3})
M mega (10^6)	μ micro (10^{-6})
k kilo (10^3)	n nano (10^{-9})
h hecto (10^2)	p pico (10^{-12})
da deca (10^1)	f femto (10^{-15})
d deci (10^{-1})	a atto (10^{-18})

Examples: nm nanometer (10^{-9} m)
kN kilonewton (10^3 N)
MeV megaelectronvolt (10^6 eV)
μW microwatt (10^{-6} W)

10. Some Extrasystem Units

Length . .	1 Å (angstrom) = 10^{-10} m
	1 AU (astronomical unit) = $1.496 \cdot 10^{11}$ m
	1 light-year = $0.946 \cdot 10^{16}$ m
	1 parsec = $3.086 \cdot 10^{16}$ m
	1 in (inch) = 0.0254 m;
	1 ft (foot) = 0.3048 m; 1 yd (yard) = 0.9144 m
	1 mile = 1609 m
Time . . .	1 day = 86400 s
	1 year = $3.11 \cdot 10^{7}$ s
Mass . . .	1 amu = $1.66 \cdot 10^{-27}$ kg; 1 oz (avdp) = 0.028 kg
	1 ton = 10^{3} kg; 1 lb (avdp) = 0.454 kg
Force . . .	1 kgf (kilogram-force) = 9.81 N
	1 tf (ton-force) = $9.81 \cdot 10^{3}$ N
Pressure . .	1 bar = 10^{5} Pa (precisely)
	1 atm (atmosphere) = $1.01 \cdot 10^{5}$ Pa
	1 mm Hg (Torr) = 133 Pa
	1 in Hg = 3386 Pa
	1 psi (pounds per square inch) = 6895 Pa
Energy . .	1 eV = $1.60 \cdot 10^{-19}$ J
	1 Wh (Watt-hour) = $3.6 \cdot 10^{3}$ J
Power . . .	1 hp (horsepower) = 736 W

11. Astronomic Quantities

	Mass, kg	Mean radius, m	Mean orbit radius, m
Sun	$1.97 \cdot 10^{30}$	$6.95 \cdot 10^{8}$	—
Earth	$5.96 \cdot 10^{24}$	$6.37 \cdot 10^{6}$	$1.50 \cdot 10^{11}$
Moon	$7.34 \cdot 10^{22}$	$1.74 \cdot 10^{6}$	$3.84 \cdot 10^{8}$

12. Fundamental Constants

Velocity of light *in vacuo* . .	$c = \begin{cases} 2.998 \cdot 10^{8}\ \text{m/s} \\ 2.998 \cdot 10^{10}\ \text{cm/s} \end{cases}$
Gravitational constant	$\gamma = \begin{cases} 6.67 \cdot 10^{-11}\ \text{m}^3/(\text{kg}\cdot\text{s}^2) \\ 6.67 \cdot 10^{-8}\ \text{cm}^3/(\text{g}\cdot\text{s}^2) \end{cases}$
Standard free fall acceleration	$g = \begin{cases} 9.807\ \text{m/s}^2 \\ 980.7\ \text{cm/s}^2 \end{cases}$

Concluded

Avogadro constant	$N_A=$	$6.025\cdot10^{26}$ kmol^{-1} $6.025\cdot10^{23}$ mol^{-1}
Elementary charge	$e=$	$1.602\cdot10^{-19}$ C $4.80\cdot10^{-10}$ esu
Electron rest mass	$m_e=$	$0.911\cdot10^{-30}$ kg $0.911\cdot10^{-27}$ g 0.511 Mev
Electron charge to mass ratio .	$e/m_e=$	$1.76\cdot10^{11}$ C/kg $5.27.10^{17}$ esu/g
Proton rest mass	$m_p=$	$1.672\cdot10^{-27}$ kg $1.672\cdot10^{-24}$ g
Atomic mass unit	1 amu=	$1.660\cdot10^{-27}$ kg $1.660\cdot10^{-24}$ g 931.4 MeV

INDEX

Acceleration 15
  angular 24
  centripetal (normal) 21
  Coriolis 32
  free fall 58
  of a point in polar coordinates 298
  relativistic transformation of 226
  tangential 21
  transformation of 31
Aiming parameter 133
Angular momentum 147
  conservation of 156
  internal 161
  relative to an axis 15
Angular velocity 24
  summation of 29
Arc coordinate 19, 55
Axes, free 173
  of inertia, principal 173

Centre of mass (inertia) 120
  motion equation of 121
C frame 9, 123
Clock synchronization 195
Coefficient of friction 52
Collision 126
  completely inelastic 127
  endoergic 133
  exoergic 133
  head-on 128
  inelastic 133
  non-central 129
  perfectly elastic 128
  of two particles 126
  of two relativistic particles 245
Conservative system 125
Contraction of length 204
Coordinates
  arc 9, 55
  Cartesian 17
  polar 257
Coordinate system 5, 54
Couple 160

Decay of a stationary particle 247
  of a moving particle 254
Diagram of momenta, vector 130
Dilation of time 200, 219
Diminution 82
Direction cosines 18
Doppler effect 223
  radial 223
  transverse 224

Einstein's postulates 194
  principle of relativity 194
Energy
  conservation of 93, 101
  decay 247
  internal 124
  internal potential, of a system 96
  kinetic 90
  kinetic, of a solid body 171
  kinetic, of a system 98
  mechanical 90
  potential 81, 95
  potential, of a system 95
  of a relativistic particle 234
  of a rotating solid body 167
  threshold 254
  total mechanical 91
Energy conservation law 93, 101
Energy content 237
Energy diagram of nuclear reactions 252
Energy-momentum relationship 238
Energy yield of nuclear reactions 246
Equation
  of dynamics, fundamental 52
  of harmonic vibrations 67
  of moments 147
  of moments in the *C* frame 163
  of relativistic dynamics, fundamental 231

Equipotential surface 86
Equivalence principle 60
Event 195

Field
    of central forces 80
    of elastic forces 83
    of force 79
    of gravity 83
    of a point mass (charge) 83
    potential 79
    stationary 79
Field potential 88
Field strength 88
Force 46
    central 80
    centrifugal 57
    conservative 79
    Coriolis 57
    Coulomb 51
    dissipative 100
    elastic 51
    equivalent 164
    external 99
    of a field 84
    friction 52
    gravitational 50
    inertial 57, 60
    internal 99
    longitudinal 233
    Lorentz 249
    magnetic 51
    non-potential 99
    normal 55
    potential 99
    of reaction 48
    reactive 137
    resistance 52
    of sliding friction 52
    tangential 55
    transverse 232
Fundamental equation of dynamics 52
    in polar coordinates 298
    of variable mass 136
Fundamental equation of relativistic dynamics 231

Galilean relativity 43
Galilean transformation 34
General theory of relativity 61
Gradient 86
Gyroeffect 178
Gyroscope 174
    precession of 175
Gyroscopic couple (moment) 177
Gyrostatic action 178

Hooke's law 52

Increment 82
Inertness 45
Initial conditions 16
Instantaneous rotation axis 28
Interval 214
    light-like 214
    space-like 214
    time-like 214
Isotropy of space 42

Keplerian motion 298
Kinematics of a point 14
    of a solid body 21
Kinetic energy 90
    relativistic 234
    of a solid body 171

Law
    of acceleration transformation, relativistic 226
    of conservation of angular momentum 156
    of conservation of energy 93, 101
    of conservation of momentum 117
    of inertia 41
    Newton's first 41
    Newton's second 46
    Newton's third 48
    of universal gravitation 50
    of velocity transformation, relativistic 215
Laws of forces 50
Light clock 201
Local time 211
Long-range action 49
Lorentz contraction 205

Lorentz factor 241
Lorentz transformation 208
  of momentum and energy 240

Mass 45
  gravitational 50
  reduced 125
  relativistic 229
  rest 229
Mass-energy relation 236
Meshchersky equation 137
Michelson's experiment 192
Minkowski diagram 217
Moment of force 148
  relative to an axis 151
Moment of inertia 166
Momentum 47, 114
  conservation of 117
  relativistic 227, 230
  of a system 115
  total, of external forces 160

Path 14
Perfectly rigid body 12
Plane motion 26
  of a solid body 170
Potential 88
Power 78
Precession 175
Principle of equivalence 60
Proper length 205
Proper time 202

Reference frame 11
  heliocentric 42
  inertial 41
  non-inertial 42
Relativistic mechanics 12
Rod-and-tube "paradox" 222
Rotation 22
  of a solid body 38
  about a stationary axis 22, 165
Rotation axis, instantaneous 28

Simultaneity 197, 210
  relativity of 219
Space-time diagram 224
Steiner's theorem 301, 167
Superposition of fields 88
Symmetric elastic scattering 251
Symmetry of space-time 42

Time dilation 200, 219
Torque 148
Transformation
  of acceleration 31
  of velocity 31
  of velocity direction 225
Translation 22, 57

Uniformity of space 42
  of time 42

Vector
  axial 24
  displacement 14, 35
  mean velocity 15
  polar 23
Vector diagram of momenta 130
Velocity
  of a centre of mass (inertia) 120
  of light 49
  of a point 14

Weight 51
Work 74
  of Coulomb force 76
  of elastic force 75
  of gravitational force 77
World line 217
World point 217